EARTH-SHATTERING

ALSO BY BOB BERMAN

Zapped

Zoom

The Sun's Heartbeat

Shooting for the Moon

Strange Universe

Cosmic Adventure

Secrets of the Night Sky

EARTH-SHATTERING

VIOLENT SUPERNOVAS, GALACTIC EXPLOSIONS, BIOLOGICAL MAYHEM, NUCLEAR MELTDOWNS, AND OTHER HAZARDS TO LIFE IN OUR UNIVERSE

BOB BERMAN

LITTLE, BROWN AND COMPANY

New York   Boston   London

Little, Brown and Company
Hachette Book Group
1290 Avenue of the Americas, New York, NY 10104
littlebrown.com

First Edition: February 2019

Little, Brown and Company is a division of Hachette Book Group, Inc.
The Little, Brown name and logo are trademarks of Hachette Book Group, Inc.

The publisher is not responsible for websites (or their content)
that are not owned by the publisher.

The Hachette Speakers Bureau provides a wide range of authors for speaking events.
To find out more, go to hachettespeakersbureau.com or call (866) 376-6591.

ISBN 978-0-316-51135-3

Library of Congress Cataloging-in-Publication Data

Names: Berman, Bob, author.
Title: Earth-Shattering : violent supernovas, galactic explosions, biological mayhem, nuclear meltdowns, and other hazards to life in our universe / Bob Berman.
Description: New York : Little, Brown and Company, [2019] | Includes index.
Identifiers: LCCN 2018026997 | ISBN 9780316511353 (hc)
Subjects: LCSH: Big bang theory — Popular works. | Cosmology — Popular works.
Classification: LCC QB991.B54 B47 2019 | DDC 523.1/8 — dc23
LC record available at https://lccn.loc.gov/2018026997

10 9 8 7 6 5 4 3 2 1

LSC-C

Printed in the United States of America

To Ulster Publishing editors Brian Hollander and Julie O'Connor and, especially, its publisher, Geddy Sveikauskas, who trusted me enough to pay for my first Woodstock Times Night Sky *column in 1975—a weekly deadline I'm still meeting today*

Contents

EARTH-SHATTERING

Author's Note

*The day which we fear as our last is but
the birthday of eternity.*
—SENECA, CIRCA AD 48

I t's not the end of the world!"
My wife muttered that cliché for perhaps the fortieth time
this year in a kindhearted attempt to make me feel better about our
snowblower's snapped pull cord. Usually one doesn't reply to rhe-
torical comments. But I found myself saying, "Unless, by amazing
coincidence, a nuclear war begins this afternoon."

She didn't laugh. Neither of us was guilty of original thinking.
Civilizations throughout time have been obsessed with Armageddon.

We can only guess why this is so. Probably humans share a
hopeful sense of collective destiny, of events unfolding according
to some lofty blueprint. And this feeling of grandeur makes people
want to lump their fate in with some epic, planetwide denouement
rather than face the more likely reality that they'll succumb to high
cholesterol.

Whatever the reason, people's fears commonly revolve around
devastating events beyond their control. During the Q-and-A
period after I give a lecture at a college or library, audience members often express worries about Earth's poles flipping or a rogue
planet colliding with ours. The 1996 *Discover* magazine issue with
the cover story about the chances of a giant meteor slamming into
our world was one of the most popular ever.

Perhaps it's the sheer drama, plain and simple. The notion of millions of people dying at the same time—whether from an Ebola-type epidemic or nuclear terrorism—seems riveting, even while the actual likeliest threats to people remain unglamorous, like lung cancer due to smoking cigarettes.

Finally, such planetwide holocausts are not purely theoretical. Devastating catastrophes have actually happened repeatedly, and they will happen again. Learning about them, especially with the addition of all the cool-fact details that have only recently surfaced and are still not widely known, satisfies a very old need that transcends today's culture. At the same time, recently evolving global threats, not just to our species but to our entire planet, bring a new urgency to the subject.

I'll admit it—I'm one of those people who find this stuff fascinating. And I'm determined to create a factual narrative that vividly illustrates these cataclysms, past, present, and future.

PART I

CATACLYSMS IN THE HEAVENS

CHAPTER 1

CATACLYSMS 101

We're all oddly drawn to a theme that started in ancient Egypt and Greece and has perennially appeared in folklore. It's the idea of the phoenix, the creature that emerges from the ashes of a conflagration. The metaphoric implication is that if our world, or at least our culture and all we hold dear, is violently destroyed, something just as vibrant can rise from the ruins.

This has actually happened, not once or twice, but repeatedly. And not just on the local or even planetary level but on epic scales that have rattled the floorless alleyways of the cosmos. We're talking about cataclysms.

In a way, they're counterintuitive. A quick study of nature shows that all objects display inertia—a strong tendency to keep doing whatever it is they are already doing. Planets whirl around the sun, and nothing is needed to maintain the motion. Meanwhile the sun, the center of all this respectful planetary circling, carries its obedient retinue through the galaxy at 144 miles a second as it participates in the galaxy's rotation. The sun and its flock have performed nineteen such circuits since their birth, and, like a mother duck with her ducklings, Sol always returns with the same retinue in tow.

No long-lived observer patiently studying our solar system for millennia would notice anything different with each galaxy rotation.

Only if he were armed with a super-telescope and perhaps alerted to be extra-attentive would the onlooker detect a change. "Look at that third planet, the blue one," he might whisper. "This time around, at some point while it's been away on its most recent galactic orbit, most of its life was destroyed. It has undergone a cataclysm. And yet now its surface is crawling with strange new organisms. What magic! What will happen next?"

Our ageless observer might be quicker to see that with each galaxy rotation and each return, the sun itself has changed. Each time it completes its 240-million-year circuit of the Milky Way's core, going beyond the stars of Sagittarius, it returns 2.5 percent more luminous. That's just a little brighter. The change is as subtle as a 75-watt bulb becoming 77 watts, so our onlooker would have to be sharp indeed to notice this alteration. Only after about four circuits of the galaxy's core and the passage of a billion years would the brightening of the sun be obvious. Still, even after that protracted interval, the difference would be a mere 10 percent boost. The 75-watt bulb is now 83 watts. But that's enough to make the third planet too hot. In the past single circuit, all its life has vanished.

A close inspection of that planet and its near neighbors shows that Earth's large-headed *Homo sapiens* creatures have built ark-like spacecraft and abandoned their world to colonize the next planet out, the orange one, although life there is proving to be a radiation-filled struggle.

The above scenario is a reasonably accurate look ahead to a time a mere 1.1 billion years from now. The sun will indeed be 10 percent brighter, and all Earth life will indeed be destroyed by the heat, and if we humans still exist, we will surely have fled to the only planet in that anti-sunward direction that has any sort of surface. All others are gaseous and slushy and offer no place to land.

This cataclysm has already happened on our planet, several

times. In those previous solar brightenings, epic terrestrial responses followed each luminosity boost. The composition of our planet's atmosphere changed in such a perfectly appropriate reaction that it might almost seem engineered. It's as if Gaia—the name for our biosphere when viewed as a single intelligent entity whose constituent plants, animals, biomes, and natural systems function cooperatively— brilliantly reconfigured the surface to comfortably stabilize it despite the extra solar luminosity.

The record reveals other slow changes too, like Earth's orbital path mutating from a round shape to an elongated egg-shaped track and back again over a 112,000-year period and the twenty-six-millennia alterations in the direction of its axial tilt, both of which dictate the onset and ending of ice ages. And there were more alchemic tricks in nature's terrestrial handbook. Super-imposed on these enormous global alterations and other powerful if snail-paced world-shapers, such as the muscular mountain-building abilities of plate tectonics, are the spices in the saucy story line— the disastrous metamorphoses that were sudden rather than grad-ual. To Earth's inhabitants at the time, these were the cataclysms. Widespread death and disruption were their calling cards, events that seemed to be conjured by the notorious Hindu goddess Kali, who preferred to destroy worlds rather than, say, enjoy a snack.

How and when these cataclysms arose, how they changed the planet and its inhabitants, and which ones could be closer to re-materializing than you might imagine is the subject of this book.

People don't think about cataclysms too often. That could be because it's anathema to our need for safety, but perhaps it's also because the topic doesn't seem relevant to our everyday lives and thus probably requires some Paul Revere type to sound the alarm before people will pay attention.

I'll volunteer.

A cataclysm is typically an event of surprise and upheaval, and it usually descends on its victims rapidly, although a relatively slow-spreading global epidemic would also qualify (and does, as you'll see in our pandemic chapters later on). If the word *cataclysm* has an antonym, it might be *safe routine,* which is nowhere more exemplified than in modern American life, where events are largely predictable and local movement leisurely.

There's not much animation out the back window of the typical suburban house, where the panorama resembles a snapshot more than a video. In the neighborhood, leaves may stir in the breeze. Clouds lethargically mutate. Even in bustling cities, motion unfolds at an accessible human clip. Walkers, bike riders, planes above, even the traffic—nothing is too fast to follow. The buildings on one's block will be there tomorrow. The reassuring steady pace is not likely to change in the foreseeable future.

Even everyday tools and gadgets have been engineered to eliminate surprises and ensure that nothing alarming stirs up the dust. Since humans are famous for not following directions—or even reading those that accompany every new purchase—most devices have built-in idiot-proofing. A portable electric heater? The directions warn against placing it where it might tip over and ignite the carpet. But if you do that anyway? Why, then there's a safety cutoff switch.

Cars are increasingly built with blind-spot monitors and backup cameras, and designers keep asking how humans can be protected against themselves. There are very few products available to the public that have significant inherent danger. The exceptions to this rule are sufficiently rare that everyone knows what they are. Two of them are chain saws and motorcycles. With such things, a single foolish moment of inattention is enough to produce grievous injury. These items simply cannot be made idiot-proof. Other semi-dangerous products exist too, but most have been grandfathered into our twenty-first-century lives. If bicycles had never been available before and someone proposed to put them on the market today, government safety agencies would never approve them. A flimsy transportation apparatus with no air bags or seat belts meant to be used on the public roads alongside motor vehicles? Are you kidding? No way!

So protective measures keep increasing, and transportation, once one of the most dangerous activities in human endeavor, has never been safer. Even a natural calamity such as an earthquake now encounters buildings and bridges specifically designed to withstand its assault. And anyway, such dramatic natural cataclysms are rare.

All this tranquillity is, thankfully, nature's rule rather than its exception. If restlessness someday drove humans to leave this planet for a more exciting venue, they'd find most celestial destinations to be even more placid than their home world. An immortal observer on the moon's surface could wait a million years and still not perceive the slightest change in the dusty lunar terrain other than the sharp-edged inky mountain shadows that gradually lengthen and then shorten in a reliable fortnightly cycle.

The overwhelming majority of celestial acreage is inactive and will remain forever unruffled. More than 90 percent of the visible universe's current seventy billion trillion suns had non-attention-getting births and are living out their lives in a steady, predictable fashion. When their lives near the end, they do not explode but merely collapse and harmlessly fade to black, as if in a movie's final scene.

But when cosmic violence does unfold, it more than rattles the neighborhood; it changes the very fabric of the universe, even if these mega-shake-ups lie near the limit of human comprehension.

Most of us don't know much about pandemonium, and that makes sense, since we're generally interested in events and circumstances that affect us personally. If you asked people to think of a dozen arbitrary topics, few would involve frenzied life-threatening catastrophes. Their thoughts might first turn to nutrition. Vacation swimming spots. Hollywood gossip. Nature lovers might think of birds or constellations or hiking trails. Even when natural violence makes the front page, people are mostly concerned with the consequences.

A series of wildfires ravages California. The TV news plays video showing flames that are awfully fast in their hypnotic dances, but the fire itself rarely spreads faster than human walking speed. Truly rapid natural motion is something else. Its rarity alone makes it startling. It always grabs our attention, partially because it displaces air and thus produces a sudden accompanying noise, and we mammals are most attentive when multiple senses are simultaneously activated. If the speedy movement involves something small, like a mosquito's wings beating 440 times a second, the sound is diminutive. The mosquito's wings produce a whine in the musical pitch of A, which coincidentally is one of the two simultaneous notes that make up a telephone dial tone.[1]

If the fast action involves something physically large, like a mile-long bolt of lightning traveling that one mile in a mere 1/6,000 of a second, then the huge mass of displaced superheated air creates a startling 100- to 120-decibel thunderclap if the bolt is nearby. We jump at the extreme noise level but normally do not contemplate its roughly 100-hertz tone bellowing subwooferly at a low bass-clef pitch of G or A.[2]

The faster the motion and the more massive the object that suddenly changes position, the greater the violence and the more likely the event gains a nomination for inclusion in our narrative. The motion needn't involve solid things. Much dynamic activity can unfold in the myriad liquid drops in the curvy cubic-kilometer assembly known as a cumulus cloud. A cloud typically weighs a million pounds, and even if its volume were solely a gaseous mass that lacked a liquid component, that much vaporous material abruptly shifting position would produce sufficient violence to attract immediate attention.

Water, which volume for volume is 784 times heavier than air, unleashes great destruction when it rapidly shifts, as was tragically seen in the 2004 Indian Ocean tsunami that claimed a quarter of a million lives. Though we will examine such earthly violence later on, these local ferocities pale in comparison to the events and processes that produce ultrapowerful novas, exploding galaxies, supernovas, and other space-time-warping cataclysms. There is also strange, off-the-radar, newly discovered violence spawned by unlikely mechanisms like extreme magnetism. And we won't neglect the lesser but still impressive shake-ups in the Earth's own neighborhood, including the catastrophically explosive birth of our moon.

We'll unravel the workings of nuclear fusion that culminated in the hydrogen bomb; its surprising mechanisms deserve to be probed in instruction-booklet detail, especially since the bomb's creation involved nail-biting wrong turns that led to human tragedies that are still largely untold.

We will see what's behind the ongoing violence that makes four trillion neutrinos zoom through every human eyeball each and every second. And we'll learn about the ultrahigh-energy cosmic rays from distant cataclysms that are continually bombarding us and discover how they affect our health. We'll examine our planet's protective mechanisms and how they nonetheless leave us periodically susceptible to the cosmic sadism that envelops us still.

We will savor the most spectacular of all violent events, the all-time-greatest pyrotechnic displays, by-products of which were the very materials nature used to fashion our brains: the Big Bang itself.

It's usually a bad entertainment strategy to begin with your highest superlative. What could ever top it? But if I am to offer any sort of timeline, the curtain cannot rise sooner than "at the beginning." This chronology is also set in stone because if time has any independent reality (which is actually doubtful), it dates to the Big Bang.

The problem, if it can be called that, is that although the Big

Bang is popularly visualized as an unimaginably violent explosion, its nature was fundamentally different from everything that followed. An explosion is a sudden outward-rushing paroxysm. A star can explode, and so can a galaxy. In war, all manner of armaments and shells detonate. In all these cases, material is flung outward. The violence is easily ascertained and qualified by the mass and velocity of the outrushing fragments.

Damage is always intimately linked with kinetic energy, which is a fancy way of saying "the power of speed." Kinetic energy is expressed as mass multiplied by velocity squared. So speed—not the weight of the material being destroyed and not the behavior of the explosive substance—is the supreme damaging factor in an explosion.

But the Big Bang was different. There, nothing was accelerated. Rather, this unimaginable violence solely involved the *frenzied expansion of space,* of emptiness itself. How that resulted in continued wild motion with consequences felt to this day is a story that has neither a true beginning nor, yet, a decisive denouement.

CHAPTER 2

IT REALLY WAS A BIG BANG

The Big Bang was the most violent incident in the universe's history. That it was utterly different from all the other cataclysms that came afterward barely expresses its uniqueness, since the event contains contradictions even in the logic needed to grasp it.

In 1929, when the brilliant but snooty and widely disliked Chicagoan Edwin Hubble announced in his faux-British accent that the universe was expanding, he opened up an argument. Until then, most cosmologists and philosophers had believed that the universe was eternal, a condition they called "the steady state." But using California's new Hooker telescope with its eight-foot-wide green mirror made of melted wine bottles, Hubble saw that all galaxy clusters were racing away from Earth.

The farther away they were, the faster they went. It was so unexpected and bewildering that many of the astronomers may have wished that a few of the bottles had been spared. The discovery suggested a violent if confusing natal moment in the distant past.

The opposing model, which was that everything always existed and never had any sort of beginning, was science's majority view until then. And as we'll see in a bit, that might actually be true, at least on a larger level.

The new alternative—that the cosmos did indeed have a birth

of some sort and that it was abrupt rather than stately—was of course not such a new idea, since it jibed with the Genesis account in the Bible, which explained that God effortlessly created the cosmos in a mere few days.[1]

Science has never been too keen on the idea of a natal moment. It was obvious that any genesis event would immediately provoke further questions about antecedent conditions and how *they* arose. It took no genius to realize that you'd frustratingly find yourself in an inconclusive series of infinite regressions.

Nonetheless, in an 1848 essay, Edgar Allan Poe wrote about the cosmos beginning as a kind of primordial egg. And a lifetime later, a few years before Hubble's announcement, a Belgian priest and astronomer named Georges Lemaître proposed, in his hypothesis of the primeval atom, that one could perhaps trace all speeding galaxy clusters back to some sort of super-dense starting point.

Extrapolating back in time is a simple task with any explosion if it's filmed in slow motion and then the movie is played in reverse. And that's our situation, as observers, as we watch the universe expand. In the case of the entire cosmos, its rate of expansion, eventually called the Hubble constant, is easily appreciated. A good analogy is a balloon with ants scattered evenly on its surface. Each ant represents a group of galaxies. Inflate the balloon further, and each ant sees its nearest neighbor move slowly away. It observes ants on the far side of the balloon recede fastest. But no ant itself expands.

Similarly, galaxies and even clusters of these cities of suns retain their sizes and do not inflate while the gaps between the clusters continually grow. As for the rate of this expansion, a longtime holy grail for astronomers, you can easily memorize this Hubble constant (drop it in conversation at parties; it will impress your listeners no matter their vocations). It's fourteen miles per second per million light-years of distance. So you just multiply fourteen by anything's distance from Earth in millions of light-years. And now you've stated how fast everything is racing away in this epic cosmic booming.

Let's do one quick example. Say a galaxy cluster is 100 million light-years distant. It must then be moving away at a speed of 100 multiplied by 14, so 1,400 miles per second. Fast enough to cross North America in two seconds.

To ponder the past, you simply trace everything's motion backward (or deflate the balloon), and it all converges, making it obvious that everything must have originated from a single spot 13.8 billion years ago.

That place of the Big Bang wasn't anywhere in the universe. Rather, it was itself the entire universe, back when the cosmos was the size of a mustard seed. The Big Bang occurred wherever you happen to be right now, since all locations were in a single spot, which has since become everywhere.

Since the Big Bang is still banging, and at an ever increasing speed to boot, we could indeed consider it an ongoing explosion in which we are part of the hurtling debris. But as we'll see, a far more violent event was embedded within the main one, and this is what will win our special-effects Oscar, especially since it fashioned the cosmos we now see around us.

But first, let's really understand the Big Bang once and for all, since scientists are actually quite sure the universe started out astonishingly tiny, dense, and hot some 13.8 billion years ago. We'll tackle this step by step, as there are multiple aspects to consider, and each is an important piece of the puzzle.

Item one: The average density of the cosmos is one atom for every cubic inch of space. This is close to the critical density that will keep it expanding forever under its own momentum.[2] If there were more material, the stronger overall gravity would have made everything collapse long ago, before stars and planets and life could form, and that obviously didn't happen. Also, more matter would make space itself visibly curve. Instead, we observe a large-scale space-time structure that is flat, meaning distant stars are just where they should be if no distortion is altering their positions.

Conversely, if there were less material, the weaker overall gravity would show itself as a runaway inflation and a negative saddle-shaped curvature of space, neither of which is observed. So we're embedded in an oddly precise and perfectly balanced life-friendly cosmic density. This critical evenness we see today had to have been present from the beginning, since the act of inflating for billions of years would have exaggerated any discrepancy.

The first takeaway: We live in an extremely peculiar universe in which the distribution of matter hangs right on the razor's edge of allowing things to continue to unfold without any imminent reality-ending large-scale cataclysm of sudden collapse or sudden expansion into cold black emptiness. And it was this way from the get-go.

Item two: Particle accelerators like the Large Hadron Collider show how matter and energy behave in extremely hot conditions, so we know that when the cosmos was only one and a half minutes old, it would have had a temperature much hotter than the sun's core that would have created matter out of energy in a specific ratio of light elements (hydrogen, helium, and lithium) and with exactly the abundances now seen throughout the cosmos.

Item three: When the universe was a trillionth of a trillionth of a trillionth of a second old, it must have suddenly but briefly inflated far faster than light speed. This wild frenzy is the only possible explanation for why, today, the temperature and curvature of space are the same everywhere and in all directions. We'll return to this frenzied inflation shortly.

Item four: About 379,000 years after the Big Bang, the subatomic particles that had formed from the intense hot energy had cooled enough to simultaneously create ordinary atoms everywhere, mostly simple hydrogen. At this moment, the dazzling omnipresent photons of light were no longer absorbed and scattered. *Space was suddenly transparent for the first time.* The fog cleared and now light could fly freely.[3] This light is still present because, in terms of the universe, nothing ever leaves to go anywhere else. Because of the continued expansion of the cosmos, all these light waves were stretched out and weakened as they traversed space for eons, and they now emanate from the entire sky as uniformly wimpy microwaves. We call this radiance the *cosmic microwave background,* or CMB.

Item five: The impressive evenness of this glow—unvarying to one part in eighty thousand—nonetheless has slight temperature knots or bumps where stars and galaxies later formed. So the CMB's overall evenness and its tiny irregularities, called *anisotropies,* explain the eventual structure of the cosmos using the known laws of nature.

Item six: The expansion of everything, with speed increasing with distance, means that anything farther than 13.8 billion light-years from us is flying away faster than light speed. That's the vast bulk of the cosmos. It's at least 99.999 percent of everything. So we cannot observe or learn anything about virtually all of the cosmos because light emanating beyond that point will never get here. Everything we see or can ever hope to see, which is everything nearer than that, might be called the *visible universe* or the *observable universe.*

Thus the visible or observable universe has a precisely defined size with a sharp cutoff 13.8 billion light-years from Coney Island. But the actual universe, which is far larger, is of unknown dimensions. It might extend infinitely.

Review these aforementioned six items and you'll understand the Big Bang better than anyone else around you (unless you're currently living in a Caltech dorm with astrophysics grad students).

The Big Bang theory accounts for the visible universe's elements, motion, construction, density, and omnipresent CMB, those microwaves coming at us from all directions. It is specific and successful. But it is silent about one or two little details. It does not try to explain why an entire universe as small as a mustard seed abruptly materialized out of nothingness one morning. Or how, hotter and denser than anything we can imagine, it briefly inflated faster than light. This "origins" business is thus a very strange situation in which we know all the details, including temperatures, sizes, components, and dates of events, to an accuracy of several decimal points, and yet it all surrounds a logically impossible occurrence whose antecedent conditions are an utter mystery.

As we've seen, the big take-home conclusions include that business about virtually the entire universe, the realm surrounding our observable universe, being forever unknowable. But there's still more. Not yet mentioned is that in 1998, astronomers trying to pin down the exact size of the visible universe by studying the light from supernovas found that when the cosmos was half its present age, it started expanding faster and faster. It was as if rocket engines were somehow attached to all galaxies and they suddenly began firing simultaneously.

Desperate for any explanation, theorists assume that empty space itself exerted an antigravity push and was blasting the cosmos apart. Called *dark energy* or *vacuum energy,* this negative gravity presumably dominated the situation long ago and was responsible for the original Big Bang; as the cosmos has gotten more spread out, the overall diluted gravity is less able to counterbalance it, so dark energy is once again becoming dominant.

This is a big deal and may even eventually destroy the universe in the ultimate cataclysm, as we'll see in chapter 14. For now, we're still digesting the idea that empty space has an antigravity property sometimes called dark energy and sometimes called vacuum energy that has the ability to push on matter. Whatever its name, its explanation is still elusive, but it may revolve around the concept that the universe arose as an enormous quantum-mechanical vacuum fluctuation whose positive mass energy is balanced by negative gravitational potential energy. If you don't really understand this explanation, don't worry; nobody else does either, since science hasn't figured out what kind of particles or fields would generate vacuum energy of sufficient strength to create this ultrapowerful, all-pervasive antigravity quality.

Whatever it is, this antigravity dark energy is necessary to explain how galaxies can move faster than light. After all, Einstein told us that nothing can be accelerated to go faster than light, and he permitted no exceptions. But here, all those planets and stars and black holes in each galaxy have never had to speed up. In a way, they're not even moving. Rather, the empty space between galaxy clusters is inflating. This is what makes the situation different from any other explosion. It's the only reason faster-than-light galaxy motion is allowed.

And it's not subtle. This wild, ongoing, ever intensifying expansion of the visible universe, a vast explosion second to none, is so astonishing it's worth fully grokking. But appreciating it requires that one understand the units used: *light-year, cubic light-year,* and perhaps also the immensity of *trillion.* Hang in here with me; this is a worthwhile time investment.

A light-year is the distance light travels in a year, 5.88 trillion miles, which doesn't usually sound impressive because the concept of a trillion is over most people's heads. But trillion might be understood by simply counting to a trillion. What's the fastest you can count out loud without making the numbers incomprehensible? Perhaps five numbers per second. So, say you counted off five numbers a second and kept going. How long would it take to reach one trillion? Answer: Five thousand years. So if you had started counting that quickly when the Great Pyramids were being built and kept ticking off five numbers a second, day and night, since then, you'd still not quite have reached one trillion today. A trillion is *big*! So a light-year, being almost six trillion miles, is a truly large distance. From Earth to Pluto is less than a thousandth of a light-year.

Our final measurement unit is a way to express big volumes. This unit is a cubic light-year—a cube that is one light-year on each side. It's so large that if you dropped suns into it at the rate of a hundred suns a second (and remember that the sun is over a million times larger by volume than planet Earth), it would take a trillion years before your cubic light-year was full of suns.

Now that we've grasped that a cubic light-year is an almost unimaginable measure of volume and that a trillion is also bigger than most of us can fully appreciate, we can start to understand the stupendous explosion that encompasses every planet in the cosmos. Ready? The visible universe is becoming forty trillion cubic light-years larger every second.

This huge, current, explosive size increase is a remnant of the Big Bang. But it's still moving at a lethargic pace compared to the epoch of inflation. That much more explosive expansion would have been necessary in order to produce a universe that has the almost perfectly uniform temperature we now observe, since 13.8 billion years, the age of the cosmos, isn't a long enough time for the universe to have reached thermal equilibrium on its own. But even such a greater-than-light-speed expansion has nagging problems, since the Big Bang was so explosively chaotic, it's unlikely that there existed some small homogenous patch that inflation could have used to increase the uniformity everywhere.

Problems or no, the inflation idea was introduced in 1979 by theorist Alan Guth to explain the uniformity modern astrophysicists observe throughout the cosmos. He knew that only a period of faster-than-light inflation could allow communication between parts of the cosmos more widely separated than even light rays could span, communication that would let physical influences like temperature and electromagnetic radiation affect adjacent atoms. Only such an expansion could permit nature's physical processes to be smoothed and averaged across the universe so that we now observe the same temperature, flat space-time, and CMB evenness everywhere.

Inflation may explain the universe's odd uniformity, but its cause is frustratingly elusive. Its nature is that of unimaginably fierce motion. It lasted just a fleeting moment, less than a billionth of a second, but during this time, the hyper-light-speed growth cataclysmically changed everything in existence. This ultimate brutality hurled apart every bit of mass and energy.

Unlike all other explosions and cataclysms that disturbed and fashioned the cosmos for the next fourteen billion years, the "why" of this unimaginable frenzy is a mystery.

In sum, basic aspects of the Big Bang make sense, and some are even simple. They explain fundamental facts of the cosmos, like why the universe is made up of particular amounts of lightweight atoms. The Big Bang explains the bizarre glow that emanates from the sky equally in all directions. And yet the theory insists on an initial utterly baffling premise: The entire cosmos popped into existence smaller than a marble and almost infinitely dense, a bewildering and seemingly impossible condition called a singularity, and that singularity suddenly performed an astounding but brief faster-than-light explosion that scientists now, in an impressive bit of understatement, call *inflation.*

We will return to singularities later when we explore yet another form of extreme violence that our eyes cannot behold—black holes.

CHAPTER 3

THE DEATH OF COUSIN THEIA

The first stars formed 180 million years after the Big Bang. Right from the get-go, these stars were bound gravitationally with their neighbors in gargantuan, blurry, pinwheel-like arrangements that coalesced from their gaseous surroundings. One of these huge galaxies, later given the giggle-worthy name Milky Way, became the site of our solar system's genesis. Thus, nine billion years after the Big Bang, in one hydrogen-rich, pinwheel-shaped, mostly blue galaxy, came a retinue of newly minted planets, including ours.

These piping-hot worlds orbited the newborn sun, which was only about half as bright as it is today, and this hodgepodge of young, still-broiling spheres of various sizes and colors took part in the galaxy's quarter-billion-year carousel-like rotation.

Because our odyssey will often take us around our home galaxy, the setting for nearly all of our cataclysms, the term *Milky Way* will keep arising. It's not inappropriate to wonder how that odd two-word label came to designate one of the grandest things the human eye can see.

Milky Way wasn't coined in some modern moment of whimsy. Although it's directly overhead the same week trick-or-treaters are handed items with the same name, the term predated the candy bar by at least half a millennium. It has an interesting etiology.

Its story began with the sky itself, whose most dominant feature in olden days—and even today, for anyone who gets away from urban light pollution—was the dramatic creamy band that bisected the heavens. This remarkable glowing belt is sadly invisible from cities and brighter suburbs, and when it is just barely glimpsed from outer suburbs, it still isn't awesome in the traditional sense of that word. But go to any desert, rural farm, isolated dirt road, or nature sanctuary on a moonless evening from August to October, and it will take your breath away. Its brilliance—it's bright enough to cast shadows—is punctuated by countless intricate inky-black Rorschach-blot patterns, making its vastness and grandeur the centerpiece of the sky. Instantly one appreciates why the Aztec and Mayan cultures revolved their mythologies around this luminescent band and regarded it as the night's dominant feature. They also considered it the path taken by the newly departed en route to heaven.

In medieval Europe, it was called by the Latin name Via Galactica, meaning "Milk Street" or, perhaps a bit more grandly, "Boulevard of Cream." It was a perfect description, since the sky band really does resemble a brook of spilled milk, albeit one that is oddly glowing. So our modern Milky Way is simply a direct translation of Via Galactica.

Its nature remained mysterious until Galileo pointed his first telescope at it in 1610. The cranky forty-six-year-old observer wrote that the glow came merely from innumerable little stars, that there was nothing more to it than that, and he added naively that people "could now dispense with wordy debates about it," which of course didn't happen. When in 1929 Hubble proved that all the sky's spiral nebulae were separate "island universes" that resembled our own, they were called galaxies too.[1] Notice that the terms *galaxy* and *galactic* contain a form of *lac*, suggesting that the entire cosmos is one vast dairy land. That's a direct consequence of that milky Via Galactica label, derived from its appearance.

So our Earth and sun were born in this Milky Way galaxy, and their early days were not quiet. Between our planet's initial oxygen-free, methane-filled atmosphere and the cataclysmic pounding that it took as asteroids and comets kept hitting it in that Wild West era when the solar system was filled with newly condensing rocky bodies in unstable orbits, it was a disagreeable time whose only redeeming feature was that the brutal unpleasantness was witnessed by no one, since the planet at that point was a world of empty grandstands.

So during Earth's first thousand million years, no living creatures were discomfited by the occasional chaos. But then, just before the earliest life mysteriously arose, the planet received an impact that would have earned Hollywood's Best Cataclysm award. It was the ultimate mayhem—a collision with Mars.

Well, not Mars exactly. But it might as well have been, for it was a planet with the same four-thousand-mile diameter. The head-on impact with Earth at 7.5 miles a second utterly destroyed the orbiting body, which was posthumously named Theia.

Our planet itself barely escaped obliteration. As it was, the collision did more than destroy the Earth's crust, the topmost layer of which has always been life's abode. Indeed, since the twenty-first-century media termed the Fukushima nuclear power plant's meltdown a *cataclysm,* what word could begin to describe something that annihilated our planet's entire crust? The English language lacks a label to denote violence of such an unimaginable degree. And the heat and kinetic energy of the collision did much worse than that. The thickest, most major layer of our earthly sphere, the mantle, was blown to bits along with the entire body of Theia.

Molten Theia fragments mixed with Earth's, and the amalgam sank into Earth's molten core, where it resides to this day, bestowing on our world a kind of mixed-breed ancestry (or so its legal origin papers would state if our planet were to be adopted by some wealthy alien collector).

But only some of that mixed-together collision detritus sank to the middle of the smoking, sputtering wreckage that was Earth. Big chunks of both planets' bodies were blasted into space to orbit around what was left of Earth. In a time frame that was more than a month but less than a century, these hot fragments coalesced into a sizzling ball that cooled gradually. We know it now as our moon.

Thus the very first cataclysm in our neighborhood of space was literally Earth-shattering. It was not some geeky, hypothetical, historical, invisible occurrence; the evidence of it parades in front of us most nights.

How do we know this crazy violence is really how the moon was born? That's a compelling story of its own.

First, our moon's orbit is more than just a bit odd; it's unique in the entire solar system.

The other major moons, a dozen of them, all orbit around their planets' equators. Each big round moon has an orbit matching its planet's spin and tilt, revealing that it and the parent planet were born together from the same glob of protoplanetary material.

At the same time, separately, more than a hundred tiny, irregularly shaped satellites orbit their parent planets in paths aligned with the flat plane of our solar system. This tells us that the little ones were merely captured asteroids that passed close to a planet, were snagged by its irresistible gravity, and started orbiting it.

Our moon fits neither of these situations.

It's huge. Fully a quarter of the Earth's diameter, it's the fourth-largest satellite in the solar system, and it's number one in terms of the ratio of its size to its planet's size. Yet it does not circle its planet's equator, something that would have signified a simultaneous birth. It oddly ignores Earth's 23½-degree axial tilt and instead orbits it in the same plane as the flat pancake-like disk of our solar system. No other major moon does this, and any theory of our moon's birth must explain this enormous quirk. If our moon behaved like all other major satellites and circled Earth's equator, it would never be seen straight overhead in places like Hawaii and

South India and even, on rare occasions, Miami, but that's clearly not the reality. So there are direct observational consequences of our moon's unique orbit around us, even if very few in today's society are sufficiently knowledgeable sky observers to notice this strange lunar behavior.

But another big oddity must also be explained. This one requires a bit of a setup. When the 842 pounds of Apollo moon rocks were analyzed, geologists got a shock. The moon's surface wasn't made of alien material. Well, yes, there were three new unearthly minerals (one was named armalcolite to honor the *Apollo 11* crew of Armstrong, Aldrin, and Collins). But these minerals were all fashioned from unbelievably earthlike atoms.

Here's where we must digress to discuss the fascinating realm of isotopes. (Stay with me—you'll like this.) Every element's nucleus has a fixed number of protons. The nucleus of hydrogen, for instance, always has one proton; carbon has six protons, and oxygen has eight. The number of protons is what makes the element what it is. If oxygen somehow lost one of its eight protons, it would instantly become a different element, nitrogen.

And except for hydrogen, every element also has neutrons in its nucleus. And this is what can vary a bit. For example, even though the commonest form of hydrogen has no neutrons, there is a rare type of hydrogen, called deuterium, that does have a neutron. Another, even rarer kind of hydrogen, tritium, has two neutrons. These three variants are hydrogen's isotopes.[2]

The most common oxygen isotope has eight neutrons, which match its eight protons, adding up to the sixteen nucleons (or particles in the nucleus) found in each normal oxygen atom; we call this isotope oxygen-16 and write it ^{16}O. It's what you're mostly inhaling at this moment, along with the nitrogen that makes up 80 percent of air. But one in every five hundred oxygen atoms has an extra neutron, making it ^{17}O. And one in two thousand oxygen atoms has two extra neutrons — ^{18}O. You're breathing all of it, a mixture of three oxygen isotopes. This same oxygen-isotope ratio is found in every mineral that has oxygen, which is essentially all rocks. Every bit of sand or quartz is silicon dioxide, or SiO_2, a compound that's mostly oxygen. In all of them, and in every grain of sand on Earth, those oxygen atoms consist of all three isotopes, and all display the familiar earthly ratios of five hundred ^{16}O atoms to one ^{17}O atom and two thousand ^{16}O atoms to one ^{18}O atom.

Now for the juicy off-Earth stuff. When a meteorite is analyzed, scientists can tell where it came from because each celestial body has its own distinctive ratio of oxygen isotopes resulting from its distance from the sun, the strength of its magnetic field, and other factors shared by no other world. So isotope ratios are like fingerprints.

Researchers have about a dozen Martian rocks in hand and a few from the asteroid Vesta. Martian rocks always have a slightly higher ratio of ^{17}O to ^{18}O than what's found on Earth, a ratio that has been standardized by using the oxygen in pure ocean water. But rocks from the asteroid Vesta have a slightly *lower* ratio of ^{17}O to ^{18}O.

The differences in the non-earthly oxygen-isotope ratios are small but easily measurable. On Mars, the fraction of rare oxygen isotopes is three parts in ten thousand higher than what's seen on Earth, and on Vesta, it's three parts in ten thousand lower.

That takes care of our setup and introduction, and this is where things get weird. Turns out, moon rocks have the *same* oxygen-isotope ratio as Earth rocks! If there's any difference between lunar and earthly oxygen-isotope ratios, it's less than one part in fifty thousand. But if the moon is composed of a combination of Earth-mantle material and Theia stuff that was hurled up by the collision cataclysm, how can this be?

Theophilus and smaller Cyrillus, to its upper right, are two of the twenty million lunar craters caused by meteor impacts, demonstrating that the moon's tumultuous birth was not the end of its violent history. *(Bob Berman)*

The moon's rocks look like our own. The puzzle of why the moon's rocks do not show an alien oxygen-isotope ratio has kept astronomers in a tizzy for decades, even if this annoying mystery isn't usually a topic at cocktail parties.

Instead of presenting a lengthy compendium of all the theories that try to explain this, I'll offer just one: Theia was most likely born in Earth's region of space, approximately the same distance from the sun, so it developed oxygen exactly the same way Earth did. Theia was therefore more similar to Earth than any celestial body we've ever analyzed, so the moon's origin through the collision scenario still makes sense, because we'd expect its oxygen isotopes to match ours.

If this theory is true, then the death of Theia was a cataclysm in yet another way: When the devastating collision totally destroyed it, Earth lost a family member. If that Mars-size, Earth-like planet were still orbiting the sun today, it could be the twin of Earth, the place to which this planet's inhabitants could someday flee when the sun monstrously inflates to engulf Earth's orbit (although if it were really near to us, it might share Earth's jeopardy when the sun inflates).

But as with many of our cataclysms, there's a bright side, and in this case, it's a literal bright side—the creation of the single useful natural night-light we all love and one that has influenced our species at a cellular level. The moon's light on the nocturnal country-side may be 450,000 times less luminous than daylight, but for countless centuries, it provided the safety margin that let people venture out after nightfall.

No one can ignore that the human estrus cycle of twenty-nine days (formerly thought to average twenty-eight days) closely resembles the lunar synodic period of twenty-nine and a half days (the lunar synodic period is the repeat time for any lunar phase, such as one full moon to the next full moon). This might be a mere coincidence, a theory supported by the fact that of all mammals, only the opossum shares the human female's twenty-eight- or twenty-nine-day menstrual period.

However, this cycle of human fertility does have a logical explanation that could indeed embrace the moon. Back in the pre-industrial, pre-electrification centuries, people ventured out at night only during the bright times of the lunar month. After all, many predators had better eyesight than humans, and a couple who took a romantic walk at the wrong time of the month could end up being eaten. Thus, if courtship did preferentially occur during the full-moon part of each month, there would be a biological reason for human females to adapt to this lunar rhythm and for fertility to evolve into a twenty-nine-day cycle.

Recent studies show that the moon's usefulness extends far beyond facilitating romance and rhyming with *June* and *tune*. The moon stabilizes Earth's axial tilt.

This is no small benefit. Planets without substantial satellites, like Mars, are gravitationally influenced by both the sun and other major planets, and they have their spin axes slowly rise and fall dramatically. A moonless Earth would do the same. When its axis was pointed most nearly vertically, the planet would have no seasons, and this situation would last for ten thousand years. When the axis migrated beyond fifty degrees and even reached the sideways position of ninety degrees, the planet would have extreme seasons with the sun straight overhead and no respite from brutal heat for long stretches of time.

Either of those conditions would prevent most earthly life from existing.

So the moon, in addition to regulating tides, also permanently protects our planet from extreme conditions and thus facilitates and nurtures life. In short, that annihilation of Theia and the near destruction of Earth was certainly a cataclysm when it happened, but it's a story with a very happy ending.

CHAPTER 4

SPOOKY THINGS THAT WENT BANG

The twenty-first century has given many of us affordable air travel that allows easy exploration of exotic locales. That's why today's vacationing city dwellers sometimes haul their families to distant undeveloped islands or sleepy tropical cays or keys whose nights are enveloped in natural darkness.

If you take a quick glance at the skyline at any of these destinations an hour after sunset, you'll see something concealed from big-city residents: nature's dark side, its nocturnal grandeur. It's fantastic and inspirational and so different from the urban firmament's synthetic glow that it takes many by surprise: *Was this really here all along, hidden behind the billboards and streetlamps?*

Suddenly you understand what the world's agrarian cultures experienced as a nightly reality up to about a century ago. You grasp the moon's former importance and why its dark monthly phases were no time for nocturnal travel. You're sobered to realize how fully technology has relegated moonlight to complete uselessness except as a romantic stage prop or a pop-song lyric.

Yet modernity has also offered the coin's other side. When you tire of being merely amazed, you can dial the experience up to astonishment if you avail yourself of today's inexpensive backyard telescope. Bygone astro-hobbyists would often grind, polish, and coat their own telescope mirrors, a months-long project, but today, the modern amateur astronomer can go out and buy a telescope complete with a motorized drive to track targets, and for just a few days' salary. Through it, the easiest and most detailed target is still the moon.

Its relevance in this exploration of cataclysms is best seen when the moon's phase is within a few days of half—although what most people think of as the half-moon is officially termed the quarter moon. Once one has shaken off the oddity that lunar astronomers use the words *half* and *quarter* interchangeably, one has to peek at the moon for only a few seconds to realize that one is viewing the remains of an apocalypse.

At the end of the previous chapter, Earth had had a devastating collision with the Mars-size Theia, and we left the moon as a hot molten ball of coalescing fragments of both bodies. It makes sense that this early molten moon would be as smooth as lava. (I just said "smooth as lava," but visitors to Hawaii who have walked on the recent lava flows from the Kilauea volcano on the Big Island know that fresh lava isn't always smooth. Far from it. The Hawaiians have words for the different lava types, though these terms lack sufficient consonants to resemble the scientific language used by serious people. The smooth lava is called *pahoehoe*, usually pronounced "pah-hooey-hooey." There's also an extremely rough lava you wouldn't want to fall on without knee pads, and this is *aa*, the beloved Scrabble word, pronounced "ah-ah.")

But looking at the moon, one sees obvious smooth dark *pahoehoe*-like areas where lava once flowed; these are the *maria*. The singular form is *mare* (say "mah-ray"), which means "sea," and the areas were so named because early observers thought the moon had oceans. The non-maria part of the moon is heavily cratered, an indication that it was mercilessly pounded by meteors. This is one of those happy situations where potentially complex science and simple visual appearance are in perfect agreement. It reveals that, following the moon's violent birth, the age of cataclysms was far from over. Soon after the lunar surface cooled came the Late Bombardment, which actually wasn't very late in the history of Earth's neighborhood; it happened some four billion years ago.

The moon was hit by so many celestial objects that the damage still looks dramatic through a telescope now, four billion years later. This easy observing project can provide entertainment for an inexpensive date, or you can use it to wow the kids while also giving them a vivid lesson in solar system history.

Low-magnification view of first-quarter moon reveals evidence of violent impacts as well as smooth, dark lava flows that have covered the oldest craters. *(Bob Berman)*

What you're seeing through that hundred-dollar instrument is nothing less than dramatic evidence of the biggest cataclysm in the history of the 'hood, because *whatever was happening to the moon was happening to Earth at the same time.* In fact, because the Earth is four times as big as the moon and has eighty-one times as much overall gravity, it attracted far more objects and received many more hits than the moon did. So if the lunar surface is a fascinating and vivid mess, imagine what Earth went through.

As with all pursuits, the more you delve, the more fun you have. Beginning guitarists and chess players learn that the basic skills acquired early on yield fun and enthrallment a year later. So if one is willing to be a careful lunar observer and think about what's visible on the lunar surface, amazing conclusions follow. And unlike quantum theory or Mideast politics, this is an area where everything makes sense.

Take a look at the southern area of the moon (if you're using binoculars, that's the bottom part of what you see; if you're using a telescope, many of which invert images, then it's usually the uppermost area). This region has the greatest concentration of craters. But why should one region get more impacts than others? It shouldn't. Common sense says that the cataclysmic era of heavy bombardment would have rained meteors equally everywhere, all over the Earth and all over the moon.

Earth was bombarded, but we no longer see craters around us here because our world comes with built-in, no-cost, clean-the-mess mechanisms. Wind, rain, and glaciers eventually resculpt every landscape; volcanoes send lava to bury some regions; and three-fourths of our planet's surface is ocean, which hides undersea scars that themselves get a tectonic makeover every two hundred million years. Given all of those sculpting and erosive mechanisms in play, only the most recent impacts are still visible here. One good example is the Meteor Crater in Arizona (called the Barringer Crater when it was visited by this author as a kid, back in 1966). Its impact hole was made only fifty thousand years ago.

Signs of an old apocalypse. This meteor crater from fifty thousand years ago is a popular visitors' destination near Winslow, Arizona. *(NASA Earth Observatory)*

Inside this giant sixty-mile-wide lunar crater, caused by a meteor collision four billion years ago, is a series of craters from smaller, more recent impacts. Except around the full moon, when the lighting is too direct and flat, you can see through any backyard telescope that the moon's surface is the scene of ongoing mayhem. *(NASA / Goddard / Arizona State University)*

On the airless, waterless moon, old craters can be erased only by flowing lava. So any part of the lunar surface that appears craterless must be lava-covered. The main lunar seas, like the famous Sea of Tranquility, are dark, smooth lava sheets.[1]

Any budget telescope will also reveal a few craters on *top* of the smooth lava fields. These are obviously due to recent impacts, as are craters inside of other craters. Thus the history of lunar violence, including the sequence of events, is dramatically displayed for anyone willing to spend a few minutes observing Earth's nearest neighbor, whose very existence is the result of yet earlier violence.

Nights when the moon is absent, the dark starry sky immediately suggests how early cultures would have detected greater cataclysms much farther away. Unlike the sunlit skies of day, the night brings news of ancient events whose light is still en route. If a flash of novel brilliance appeared, observers couldn't help but notice it in the nighttime firmament. The new star's light required thousands of years to arrive, so the brand-new sight would paradoxically be an old image of an object that no longer exists, like a fading sepia photograph.

Plus the night can be spooky, and it's responsible for that common fear scotophobia, fear of darkness.[2]

The midnight sky is thus the perfect setting for us to peek at far-off cataclysms. And not just because the dark background highlights any newly appearing light. In virtually all early theologies, the sky represented the abode of the gods, a place where stability ruled. In contrast to the whirlwinds of earthly life, from the crashing of waves to frightening thunderstorms, the night skies offered a welcome constancy, and if this was disturbed, the alteration did not go unnoticed. A new light or a significant nocturnal change wasn't merely interesting; it was weird and spooky, or even terrifying.

Constellations displayed the same patterns year after year, and even the seven bright objects that moved in front of them — the sun, moon, and five planets—kept a predictable schedule. No one knew why one of them was so much slower than the others, but this single lethargic point of light, Saturn, took nearly thirty years to move through all the zodiacal constellations and return from whence it started. Jupiter's circuit required a dozen years. The other planets circled the sky in under three years; the sun did it in one year, and the moon took a mere month.

Except for the occasional meteor flashing ephemerally, it was all very predictable. Even the brightness of these objects could be forecast. The light of Mars varied by an enormous amount, and Mercury altered its light a thousandfold during the course of a year. But Saturn and Jupiter brightened and dimmed only a little.

Most of the background stars stayed steady, although a few varied. One autumn star lost half its light every two days, twenty hours, and forty-nine minutes, then slowly got it back after ten hours. This luminosity weirdness, which repeated in a predictable pattern, was duly noted; the Arabs in the desert called it Algol, which meant "the Ghoul" or "the Demon Star." The precision of those changes tormented star watchers; it took centuries, and the invention of spectroscopy, for scientists to learn that Algol was a binary system whose orbital plane angled sideways toward the Earth, so one star eclipsed the other in an exact orbital period.

A few times in the average person's lifetime, the most demonic celestial body of all entered the scene: a comet. Here was a major strange-looking apparition with an inconstant tail that slightly altered its brightness and position from night to night. (Comets never zoom across the sky like meteors, although the public often conflates the two.) Comets violated the security and constancy of the heavens and were always regarded as portents of evil.

When the rational Greeks started to try to make sense of things, they tackled the night sky first. The Greek philosopher who was the farthest ahead of the rest of the world is, paradoxically, the one who is least well known today: Aristarchus, who lived on the island of Samos. Your author traveled to his island in 2012 on a quest to learn more about him and was initially hopeful to see that the island's airport bore his name rather than that of its more famous citizen Pythagoras. But no local library or historic site possessed any real knowledge about his life, which remains as steeped in mystery at his birthplace as it is everywhere else. This we know: Aristarchus was the first to say that our world spins on an axis while circling around the sun. He was thus eighteen hundred years ahead of Copernicus in advocating heliocentrism, even though almost everyone these days wrongly credits Copernicus or Galileo as being the first to say that the sun is stationary.

Aristarchus also figured the rough distances to the moon and sun, and these were no small bursts of astronomical enlightenment. He wanted to tackle the distance to the stars too, but that presented too much of a challenge. A century later, the Greek Hipparchus created a map of eight hundred and fifty stars, and some four hundred years later, the Greek Ptolemy added one hundred and seventy more to the chart. These maps plotting over a thousand stars proved invaluable right into the Renaissance, since a better star atlas wasn't created until 1725. Ptolemy even went a step further and added exact brightness labels. His system has been used ever since, even by modern astrophysicists, and since most stellar cataclysms are recognized by us distant onlookers solely by their changes in brilliance, we must speak the brightness language of astronomy, and that means we will use the word *magnitude.*

Ptolemy's magnitude system is easy. He called the most brilliant stars *first magnitude* and those a bit less bright *second magnitude.* The numbers kept stepping down to the sixth magnitude, which are the faintest objects the naked eye can see.

When modern scientific methods entered the scene during the Renaissance, especially the discipline of photometry (measuring light), researchers found that they could retain Ptolemy's magnitude business if they added a few simple rules.

Sixth magnitude would still be the faintest stars. The eye can distinguish only geometric brightness changes, like luminosity doublings, so each of Ptolemy's magnitudes roughly indicated a brightness doubling. To be exact, a five-magnitude change, such as the difference between a first-magnitude star and one of sixth magnitude, is a hundredfold change in brightness.

We're almost done, and this is worthwhile because the brightness stuff is going to come back a bit later to amaze us. Renaissance astronomers also realized that a few stars and planets were much brighter than all others. When the intellectual smoke cleared, some famous bright stars, like Antares and Regulus, stayed categorized as first magnitude, just as they'd been two millennia earlier; they are among the classic first-magnitude stars beloved by the ancient Greeks and desert-dwelling Arabs. But a handful of stars, like Vega and Arcturus, are even brighter and thus are categorized as magnitude 0. And one winter star is so very bright that it bursts decisively into negative numbers. This is the Dog Star, Sirius, at magnitude –1.5. Planets do even better. Jupiter usually shines at around magnitude –2.5 and Venus gets as dazzling as magnitude –5.

At its brightest, Venus can cast shadows, although in modern times the artificial glow from cities and streetlamps masks the phenomenon; you have to go to low-population-density places where the pizza is usually bad to see it. Still, think of anything that hits magnitude –5 or brighter as something that truly grabs attention in the sky and, in rural areas, might even light up the ground. The full moon reaches magnitude –12.6.

The ancient Greeks were replaced by the more brutish Roman Empire around AD 300, but they left behind a legacy of catalogs pinpointing star locations and magnitude readings detailing stellar brightness. From then onward, observers could look at the sky, compare what they saw with a reference, and see whether anything had changed.

Few if any Europeans knew it back then, but records of the

heavens actually already existed. In a very different part of the world, sky observers had been carefully chronicling changes in the firmament starting around 200 BC. These were the Chinese.

It is interesting that some early major civilizations that possessed highly organized cultures and written languages, like the Hebrews, showed no interest in keeping track of celestial events, while a few others obsessively chronicled the sky. It may be because the former's religious paradigm had no place for celestial influence, whereas the Chinese saw connectedness between Earth and heaven.

Each Chinese emperor had official court astronomers, observatories to which they were assigned to watch the sky, and a kind of sky council that advised him of anything special that appeared. The astronomers also maintained calendars. It was serious business, and later researchers benefited from their accurate, ongoing records of extraordinary explosive events in our sector of the galaxy—or at least, they benefited from those records that started after about 200 BC, which was when emperors stopped burning the records of their predecessors, and chronicles remained extant and consistent.

For example, we have a report from the later Han dynasty of a total solar eclipse on December 16, AD 65. Using a modern computer program, your author easily verified this and confirmed a one-and-a-half-minute path of totality directly over northeastern China on that date. Such cross-checks of reliability give us confidence that the rare reports of astoundingly brilliant "guest stars" can indeed be trusted, particularly when today's optical and radio telescopes find telltale celestial remnants in the same location.

But like Sherlock Holmes's dog that didn't bark, long periods of Chinese silence are just as intriguing. Truly epic star- and planet-destroying events in our own galaxy, particularly within about six hundred light-years, would definitely appear in the sky as a sudden new star of great naked-eye brightness, meaning magnitude –5 or better. A nearby one ought to be dazzling, perhaps even shadow-casting. A distant one, especially if it lay behind nebulous dust clouds, might barely be seen by the naked eye as magnitude 5 or 6. But since the Chinese court astronomers were watching for anything weird whatsoever and dutifully noted all changes in the heavens, the many long gaps in their guest-star chronicles tells us that supernovas are rare occurrences.

Nowadays we know that one or two supernovas per century in any given galaxy, including ours, is about the normal frequency. Given that there are about four hundred billion stars in a typical major galaxy, that means that only one out of every forty billion stars will go supernova during each millennium. It is a rare event.

So when astronomers reassure worriers that their vacation plans will not be spoiled by the sun exploding, the scientists are using the logic of astrophysics, which says that the only suns that blow up are unusually massive ones or white-dwarf members of a double-star system—neither of which applies to our sun. But astronomers can also offer reassurance based on statistics: In any given thousand-year span, the odds of a random, generic star blowing up are just one in forty billion: a thousand times less likely than a person winning a mega-lottery.

This is why scientists are not surprised that no one has seen a Venus-bright supernova in several centuries. It also means that when it does happen, people on Earth are witnessing not just cataclysmic violence but a relatively nearby event of extreme rarity.

And now we've set the stage for rarities within rarities, violence that truly defied the odds. These events took place within just one millennium, from the year 1000 to the year 2000. Two nearby cataclysmic supernovas *happened in the span of a single lifetime*. And then this double whammy occurred *again*.

Impossible? Not so much. We will soon see that in 1006, a nearby sun exploded with enough violence to create a brilliant, shadow-casting new star; forty-eight years later, in 1054, another star did the same thing. Then in 1572, after half a millennium of calm, a supernova dazzled the heavens; thirty-two years later, in 1604, another one appeared.

There's been nothing since.

Two pairs of dazzling supernovas. Four cataclysms that changed our galaxy. The first duo happened when observers could do no more than stare in fear and perhaps note where in the sky the weirdness was unfolding. But the second pair was a different story. This specific pair of titanic explosions were destined to change our world.

CHAPTER 5

BLAME IT ON THE SUPERNOVA

Big changes happened in the eleventh century and nobody was happy about them.

This was not a glamorous period of history, yet it was the setting for the kind of mega-blast cataclysm that nature sadistically reserves for special occasions.

Europe was entering the High Middle Ages, but as far as we know, no one was getting high. It was a depressing period now referred to as the Dark Ages, because the sophisticated architectural advances of the Roman Empire and the intellectual reasoning popularized by the Greeks lay a long, gloomy millennium in the past. Everyday life was hard and short. Periodic famines sometimes endured for years. Even the solace of the afterlife offered by the Church came with news of a great schism between Christianity's Western and Eastern branches, a schism that would last for a thousand years.

But things were different in the Middle East and China. There, science and technology were at their zeniths. China's Song dynasty actively encouraged discovery, while Arabia's Alhazen published breakthrough findings on the nature of light centuries ahead of Newton. During this Islamic golden age, stable political structures were established, trade flourished, and commerce between the Arabian lands and China thrived via ships and along overland routes between the two societies. Together, these civilizations constituted Earth's brightest light during the eleventh century.

It was then that our galaxy produced the biggest explosion ever known. And then, a mere forty-eight years later, the second-biggest.

The first of the eleventh-century blasts was the largest on record at the time, while the second became the most famous. Nobody back then would have called what he or she was seeing a *supernova*. The word *nova* first appeared after a Danish astronomer observed the months-long pinpoint brilliance in Cassiopeia in 1572 and wrote it up in his small book *De nova stella* (Latin for "concerning the new star").

The new stars seen in the eleventh century were indeed "new" in that they hadn't previously been witnessed in the ancient star patterns. Astronomers eventually learned that novas happened only to old stars, yet *new* still manages to be an appropriate adjective because new elements were created by both of these eleventh-century blasts. This act of minting a new element out of old ones, that hopeless goal of alchemists for so many centuries, serves vital functions in a variety of ways, including aiding in the formation of life, and we'll explore these supernova concoctions a bit later. But first let's imagine ourselves living back then, a millennium ago.

Written records from China and the Middle East show that new stars had been seen in the post–Roman Empire years of 185, 386, and 393. But the first of these is thought to have been a comet, while the other two never equaled the brilliance of the night's brightest stars. No such ambiguity exists when it comes to the extraordinary happening on April 30, 1006.

Times were difficult. The average life span of the three hundred million humans on the planet was around fifty years, and that was if they managed to survive childhood and various ailments ranging from malaria to ergotism (caused by a fungus on grains). Societies were either agrarian or nomadic, so people were outdoors nearly all the time. There was, of course, no light pollution, and the clarity of the night sky would be difficult for most of us to imagine. The Milky Way cast shadows.

The sky was important because it alone offered vital sources of light by day and night and because predictable changes in the positions and motions of the moon and sun (and, for the Maya, Venus) dictated the seasons and, consequently, the times for planting or harvesting. A new sky object was a big event, and not a particularly welcome one. The constancy of the starry firmament was one of the few aspects of life that offered stability and safety; moreover, priests and holy books in all cultures had made the celestial realm the abode of the gods (or the one God). When something changed— for example, when a bright comet arrived, which happened an average of once every fifteen to twenty years—it was a deeply significant omen. In no civilization did a new object in the sky mean anything good.

At least with comets, the elders of one's city or tribe had had prior experience with such an apparition and had obviously survived it. In the ever-enjoyable game of one-upmanship, some village or tribal seniors probably used the occasion to reminisce. "Yes, that's a comet, I suppose, but you think *that* looks bright? You should have seen the one that appeared when I was little, in the days of our former chief. Now, that *really* freaked us out!"

But a new star was something else. It was not an occasion for reminiscence but for wariness and insecurity. Stars were constant. Their patterns were unchanging, except for the five "wanderers" (planets) that everyone recognized, and no new stars had appeared in living memory of any townspeople, including the high priests. Yet here at the end of April 1006, low in the south sky, was a new star that was bright enough to make trees cast shadows.

Though we now know that the star that exploded lay a whopping 7,200 light-years away—just slightly farther than the most distant of the night's visible-to-the-naked-eye stars—it still happened in the nearest 6 percent of our galaxy and thus attained a brilliance in the night sky never seen before or since. This was the brightest starlike object in recorded history.

It drew excited mentions in records from Europe, Egypt, Japan, and China and probably was responsible for some petroglyphs in the Americas. Among the most extensive notes were those written by Egyptian astronomer Ali ibn Ridwan, who said that the night sky was "shining because of its light" and judged its brilliance as greater than that of Venus and perhaps one-fourth as great as the full moon. We now deem that it was magnitude –7, the greatest brilliance ever seen in a starlike point.

Others confirmed that it cast shadows during the night and was clearly visible during the day. The only sour note for some observers was that this sudden intense brilliance occurred in the far southern constellation of Lupus the Wolf, so in northern temperate places like China and Switzerland, the supernova barely cleared the southern horizon even when it reached its highest nightly position. Also, as seen from such northern locations, the new star rose, ascended a mere degree, and then set within an hour. The slightest bit of southern foliage or topographical hilliness would have hidden it. Observers in Egypt or Mesoamerica, however, would have seen it prominently.

The actual light from this titanic explosion equaled the brilliance of six billion suns, though of course our planet intercepted only a tiny fraction of that. It maintained much of its hyper-Venus dazzle for three months and then slowly faded, but it could still be glimpsed a full year later. And it wasn't just its visible light that found its way to our planet. This kind of supernova emits intense gamma rays, the most energetic form of electromagnetic radiation, and these can degrade Earth's protective ozone layer and affect the biosphere. Sure enough, studies of Antarctic cores show that an ice layer that dates from that period contains bizarre nitrate deposits that must have been caused by that explosion's photons.

The supersize telescopes of the twentieth century found a faint remnant of this blast, and the emissions from this small, ragged detritus suggest that this was a type 1 supernova, which typically attains the greatest brilliancies of exploding stars.

A Danish astronomer used the word *nova* to describe the sudden appearance of a Cassiopeia star, and the term was then used for centuries to refer to any newly appearing star, no matter how dim, and even for the sudden flare-up of an existing star. The word *supernova* didn't appear until the twentieth century, when it was coined by the genius theoretician Fritz Zwicky during a lecture at Caltech in 1931. The prefix *super-* was well deserved. An ordinary nova is typically caused by atmospheric material from one star being gravitationally captured and deposited like accumulating snow on the surface of a dense orbiting companion, a commonplace white dwarf. This hydrogen gas is gravitationally pulled down and compressed into a perilously thin layer. Eventually a critical amount of material collects, like TNT waiting for a lit fuse. The white heat of the star just beneath it provides the match. The layer of deposited star-stuff is now ready to rock and roll. When it ignites in a star-wide fusion reaction, it emits the brilliance of anywhere from five to thirty thousand suns. This is a nova.

The star is not destroyed, and the accretion process can begin again, producing periodic nova eruptions. But Zwicky was trying to get his audience to wrap their heads around the cases where the entire white dwarf's body is involved in the fusion process. If the whole star, not just the newly deposited surface detritus, goes *ka-boom,* the brilliance is typically a million times greater. In a word, *super.*

Later astrophysicists would label that process of mass accumulation from one star to a companion followed by a star-destroying blast as a type 1 supernova. They also found that whenever that occurs, it creates about the same level of brilliance, which also happens to be the greatest point-like intensity of anything in the cosmos except for quasars. They realized that having such known-intensity lightbulbs appearing periodically throughout the cosmos—once or twice a century in each and every one of the universe's one trillion visible galaxies—offered an easy way to ascertain distances and therefore the size of the cosmos.

After all, if you know the brightness of a 60-watt incandescent bulb (800 lumens) and you know your neighbor owns only that particular light, having seen him buy a box of five hundred at a giant warehouse sale, you could use a photometer and measure how bright each bulb appears as you look out your window. Then you could calculate your home's distance from your neighbor's simply by using the inverse-square law.

Wait, really, this is easy. The law states that if you double the distance you are from a light, it will appear four times dimmer (2 x 2 = 4). If you move ten times closer, it'll seem a hundred times brighter (10 x 10 = 100).[1]

In other words, we can look around the universe, find a type 1 supernova in some galaxy or another, measure its brightness, compare it with how bright such supernovas appear at a known standard distance, and then calculate how far away the newly observed supernova and its host galaxy must be. Do this often enough and you get a good sampling of true distances to galaxies. Eventually you grok the correct dimensions of the entire visible universe.[2] And all because subatomic pieces from one star are pulled onto the surface of a tiny rock-hard companion, where they accumulate until it all blows up as if it's Odin's birthday party, with the power of a billion trillion hydrogen bombs.

And this is exactly what happened that late April evening in 1006. But it's not just its brightness that makes a supernova worthy of attention, nor is it the cataclysmic destruction of an entire celestial neighborhood. The fantastic temperatures fuse together atoms to create heavy elements like iodine, gold, lead, and uranium. There's no other way to produce these. So the indispensable iodine used by your thyroid gland came solely from supernovas. Five billion years ago, the detritus of a supernova mixed with an ordinary hydrogen gas cloud, a nebula, to create the substances that formed our sun and its retinue of planets, with enough supernova residue left over to be present in your body right now.

How the universe came to have these heavy elements was one of the top five scientific puzzles a century ago. In 1920, the brilliant British astronomer Arthur Eddington finally showed how the sun shines: by fusing together four atoms of hydrogen into one of helium in its core, which released energy according to Einstein's newly minted relativity theories. From there, scientists could figure out why the cosmos has ninety-two natural elements, every single one of which is found here on Earth.[3]

As we saw in chapter 2, the Big Bang's heat created hydrogen, helium, and lithium. But now physicists could understand how the stellar fusion process could cause stars to fuse hydrogen to helium, and then helium to carbon and oxygen, and then those atoms to other atoms to create more massive atoms. It also explained why very old stars that weighed about the same as our sun had used up their hydrogen and even helium fuel, then imploded and become tiny Earth-size stars composed solely of carbon and oxygen, called white dwarfs. Slowly, as astronomers observed white dwarfs all around in space, they realized that this was the end point in the life cycles of countless trillions of ordinary sun-like stars.

Out of fuel and down on its luck, the white dwarf, an aged ball, radiates intense leftover remnant white light but cannot manufacture more energy. It has no fusion fuel remaining. If it adjusts itself slightly and collapses an additional few inches, its gravitational energy is converted to the kinetic energy of heat, and this bit of periodic shrinkage keeps the star shining for another few million years.

A sun-mass star should reach this stage about nine billion years after its birth. Thus, by now, thirteen billion years after the first stars were born, there ought to be numerous white dwarfs finishing out their lives in every galaxy.

And indeed there are. Right here in our own neighborhood of the Milky Way, eight white dwarfs are among the one hundred stars nearest to Earth. In fact, the eighth-nearest star, the famous companion of Sirius, the Dog Star, is a white dwarf, and it's just eight and a half light-years away. Another, Omicron Eridani, to the lower right of Orion's famous foot star, Rigel, can be glimpsed with ordinary binoculars.

White dwarfs are so crushed, a teaspoonful of their material would outweigh a cement truck. Every gallon of a white dwarf's body weighs more than a loaded freight train. Its surface may technically be a gas, but that gas is so dense that it's solid, even if the dwarf's white-hot starlike temperature would prohibit any astronaut or space probe from ever landing on it. No wonder its gravity is intense enough to capture the stellar wind from a companion star.

What happens next — specifically, whether this captured gas ignites to produce a nova or whether detonation is delayed until it's a star-destroying supernova — depends on several factors that include the star's mass.

A white dwarf typically weighs the same as our sun, and the great Indian astrophysicist Subrahmanyan Chandrasekhar once explained to this author why the now-famous weight cutoff, commonly called the Chandrasekhar limit, is 1.44 times the sun's mass. It's because at that mass limit, the compression of the star's core creates temperatures and other conditions that force a star-wide fusion explosion. If you reach that kind of weight, you must explode, so you can't be a white dwarf and weigh more than around one and a half suns. Do you remember why we are carrying on about this? It's because this is the only mechanism that can create a type 1 supernova, which always involves a white dwarf in a binary-star system. Thus, neither the most common stars, which are all much lighter than our sun, nor the very rare, truly massive stars, which start out weighing eight or twelve or even twenty suns, will ever collapse into white dwarfs, and they won't be the agents or victims in a type 1 supernova. Yes, this may be a tad technical, but hey, when you reach the end of this chapter, you will be able to tell everyone the difference between the two kinds of supernovas that have been periodically reducing property values in galaxies everywhere, including our own.

So, to review, a type 1 supernova happens when material from a star gets pulled onto a white-dwarf companion. It's cataclysmic, although the slightly less explosive type 2 supernova isn't much gentler, since that neighborhood-destroying cataclysm is famous for producing floods of lethal gamma rays. If either variety went off anywhere within thirty light-years of Earth — which happens statistically once every 240 million years — then it is mass-extinction time for our overworked biosphere. At a minimum, it's a terrestrial season of increased mutations, evolutionary changes, and local tumors.

A type 2 supernova does not require material from one star to accrete onto the surface of another. It's a solo event. But its blast will still dependably destroy the entire solar system and all life on any planets. Its gamma rays can sterilize other stars' planets too. It's bad news for the whole neighborhood. But it, too, ends up creating heavier elements that enrich the interstellar medium and make for a more interesting galaxy.

Here's the type 2 supernova scenario. Start with a massive star, meaning one weighing at least eight suns. These are rare and constitute less than 1 percent of the stars in the galaxy. Because the core pressure in such stars is so high, the star's center has an extraordinarily rapid fusion system that makes it superhot and causes it to burn through its fuel rapidly.

Such stars are typically red in their early years and in their old age but blue-hot in their salad days of optimum fuel-burning. We have no stars like that in our section of the galaxy. The most massive star around here, 8.5 light-years away, is the Dog Star, Sirius, which weighs "just" 2.4 times more than our own sun. That's enough to make it blue-hot, or at least a pale blue like a diamond, but it's not heavy enough that it will ever go supernova.

But let's look five hundred light-years away to the famous red giant Betelgeuse marking Orion's left shoulder (well, it's to our left as we face it, but if you were Orion, it'd be your right shoulder). Now, *there's* a superheavy star, probably weighing as much as a dozen suns. It's in its old age. It could go supernova any day now. Maybe it has, and the blinding brilliance, along with those unwanted gamma rays, is already en route.

In normal stars, hydrogen fuses into helium. When the hydrogen is almost used up, the star's light comes increasingly from other, less efficient fusion processes, such as helium fusing into carbon and oxygen. Eventually there's too little fusion going on in the core, and this spells trouble with a capital *S*.

A massive star has a lot of gravity, so its outer layers always press down heavily on its core. But when the outward-pushing core energy counterbalances the weight, the star remains stable in size. This equilibrium explains why our sun stays the same exact size, thankfully.

In older massive stars, signs of instability start appearing. Betelgeuse and most other red supergiants vary their light irregularly as they get larger and smaller in a worrisome imbalance that has got to be driving any intelligent life-forms on its planets to seek counseling.

The dying core emits too little oomph to hold back so much weight. The star's huge mass starts to collapse. But this implosion creates greater core pressure, which then ignites faster fuel-burning, and this newly released energy pushes back on the collapse. That not only halts the implosion but pushes the star to expand. It's not a subtle yoyo. In a roughly three-year period, Betelgeuse grows hugely larger and then shrinks back down again. It goes from a medium sphere somewhat larger than the size of Earth's orbit to a gargantuan ball about the size of Jupiter's orbit.

But someday there won't be enough core fuel to push back, and then the collapse will keep going. As it does so, it greatly heats up until the entire star's weight hits the superhigh-density core and rebounds back, with this shock wave igniting everything. The all-star blowup is called a type 2 supernova.

Beyond the difference in brilliance, we can distinguish a type 1 supernova from a type 2 by the light curve—that is, by measuring the supernova's maximum brightness and noting how quickly it fades. A type 1 can equal the light of over a billion suns, which makes it ten times brighter than a type 2. More distinctively, a spectrographic study shows the presence of hydrogen in a type 2 but not in a type 1.

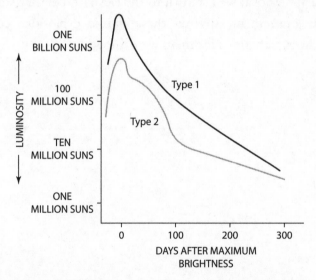

Type 1 supernovas are brighter than type 2s. But while the light curve is useful, a spectrographic study decisively reveals the supernova's nature. For example, unlike a type 1 supernova, a type 2 shows the signature of hydrogen.

Another very interesting effect characterizes a type 2. Here, much of the light is caused by the extreme heat from newly minted, intensely radioactive materials. The erupted contents of a massive star's core, hurled into space, creates lots of radioactive nickel-56.

Its half-life of just one week makes it wildly "hot" as it decays into cobalt-56. This in turn decays in seventy-seven days into the stable metal iron-56. The type 2 supernova's slow visual fading exactly parallels the time frame for the two-step decays of these radioactive materials. Eventually, ordinary iron is left flying rapidly through space. Some such iron ended up in the bodies of the newborn sun and in our own Earth, where it makes up much of the planet's core and all of the iron in your blood's hemoglobin.

Thus the catastrophic destruction of stars, planets, and galactic neighborhoods, like most other cataclysms, creates more good than harm. With all this technical stuff out of the way, and now that you understand the very different events that cause a type 1 versus a type 2 supernova, we'll return to the eleventh-century stargazers who, in amazement, spotted these titanic explosions centuries before the explanations for them were known.

CHAPTER 6

ARMAGEDDON MONUMENT

After the new-star sighting in April 1006, forty-eight years passed during which nothing momentous occurred. Earthly history was spartan. Nobody was inventing things like printing presses. Few explorers were heading out to sea to find new lands. The sky alone offered noteworthy if disconcerting changes. The most-watched, highest-Nielsen-rating worrisome event happened later in the century, in 1066, when Halley's comet made an unusually

A small section of the Bayeux Tapestry depicting Halley's comet in 1066. The enormous 230-foot-long strip of linen was actually an embroidery containing 1,515 objects, commemorating the defeat of the English Saxons by the Normans. *(Creative Commons)*

bright appearance. A comet was assumed to signify something disastrous coming—in this case, Halley's comet accompanied the defeat of the English Saxons in the Battle of Hastings.

But a mere dozen years earlier, a far more cataclysmic event had unfolded in the heavens, and it spooked everyone just as thoroughly as the later comet did. On July 4, 1054, a massive blue star weighing as much as ten of our suns used up the last of its nuclear fuel. Its struggle against its own enormous inward-pulling gravity came to an end. Its core no longer pushed outward with enough nuclear oomph to stop its upper layers from collapsing inward. Stars are gaseous, and gas easily compresses, so once it began, the implosion turned into a runaway.

The smaller a star gets, the greater the gravitational pull at its surface; that makes it still smaller, and on and on it goes. This remains nature's most astonishing vicious circle. It's certainly the most violent self-amplifying snowball effect in the cosmos.

Anything that falls converts its gravitational potential energy to energy of some other kind, like a falling boulder gaining speed and becoming more destructive. In this case, the conversion created ultra-extreme heat. In less than a minute, the star's new super-high temperature made an unusual kind of nuclear fusion unfold, quite different from the way our sun shines. The entire massive body of the huge but shrinking blue sun ignited. In mere seconds, the star grew a billion times brighter. It had become a supernova, a type 2, and *this* particular unfortunate star became astronomers' all-time-favorite hyper-cataclysm.

Two things happened at once during that single minute a millennium ago. First, most of the top gaseous layers of the unfortunate blue sun exploded outward with the total brightness of a billion suns. This destroyed all the planets orbiting it. Simultaneously, the material that had already fallen inward kept going, accelerated by gravity toward the center, until it formed a ball few physicists really believed could exist until it was discovered in 1967.

But before we examine the astonishing results of this supernova, let's imagine ourselves back in July of 1054 among the mostly agrarian humans then alive, every one of whom surely stared at the sky in disbelief.

The most impressive scene appeared at the outset, just before dawn on July 5. There, in the wee hours, a thin waning crescent moon rose in the east-northeast in the constellation of Taurus the Bull. But what on earth was next to it?

It looked like a dazzling new star. But it couldn't be. Though it twinkled madly, which planets rarely do, it was far brighter than any star or planet. The object's brilliance equaled the moon's except that all its light was contained in a tiny dot with emanating spikes or rays, the result of the normal astigmatism every eye possesses. It lit up the countryside everywhere. Early risers that morning could see their shadows cast by the dazzling new light. At magnitude −6, it was brighter than Venus at its best. And it was only slightly less intense than the 1006 supernova that had alarmed everyone forty-eight years earlier—and this one was much higher up in the heavens. Such brilliance in a starlike point has never been seen since.

Morning twilight began less than an hour after the new star rose. But the blazing object remained clearly visible in the blue sky even after sunrise. Indeed, it persisted as a daytime star for another three weeks. And though it slowly faded thereafter, it was visible at night for a full year, glued to that spot in Taurus near a medium-bright star astronomers would later name Zeta Tauri, which marks the Bull's left horn. Well, it's to our left as we face it. If you were the Bull, it would be your right horn. (After this, we'll always assume that left means the observer's left, not the constellation's.)

If the impossibly bright new star had appeared fifteen hundred years earlier, the inquisitive Greeks would have written about it and speculated about its nature. But in eleventh-century Europe, the reaction was very different. Although most clergymen and some laypeople were literate, not a single writer made mention of this dazzling phenomenon, even though it dominated the night for over a year. The silence was total. And the reason for this is obvious: The sky and its contents belonged to the "heavenly realm" that, according to Church doctrine, was immutable. The unchanging aspect of the cosmos was gospel. Since alterations in the heavens were irreconcilable with the prevailing theology, the strange new object was awkward and unwelcome. The predawn sky was essentially shouting, *The Church is wrong!* Throughout Europe, not one chronicle mentions a single word about this cosmic faux pas.

Things, however, were different in China and the Middle East. In both places, scribes penned detailed accounts of this guest star, this visitor. As one example, in the *Song Shi,* the official annals of the Song dynasty, an entry in chapter 12 dated April 6, 1056, says, "The director of the Office of Astronomy reported during the fifth lunar month of the first year of the Zhihe era, a guest star had appeared at dawn, in the direction of the east… now it has disappeared."

Here and there in primitive parts of the world, people created paintings of it.

Of course, acknowledging the star was one thing; people didn't attempt to explain what they were witnessing. It would have seemed a narrative as coherent as a cat's dreams. Knowledge of stellar distances lay half a millennium in the future, and even if they had somehow possessed such information, they still would have been flummoxed to learn that this new luminary lay far beyond the supposedly farthest objects, the night's stars.

A typical naked-eye star lies a few hundred light-years from Earth, though some of our neighbors are ten times closer. The most remote—extraordinary beacons that manage to cast their light across vast distances—hover 3,500 light-years away. But this new star in Taurus sits 6,500 light-years in the distance.

A petroglyph in Chaco Canyon, New Mexico, made by the Anasazi who lived there. Scholars say it depicts the 1054 supernova in Taurus, as the handprint accurately represents the position of the new star relative to the crescent moon that was right in that same spot that morning and to the nearby star Zeta Tauri. *(Anasazi Photography)*

Thus the event's light had taken that long to get here. This explosion had been en route to Earth since before the Great Pyramids were built.

No one on eleventh-century Earth could even suspect what was happening, but today we can use the hard-won knowledge acquired since then to confidently visualize the unfolding tragedy. It was a schizophrenic tale of huge quantities of brilliant star-material heading opposite ways, each destined to create something so bizarre, they remain monuments to nature's penchant for inventing seemingly impossible structures.

The bulk of the star furiously exploded outward with a brilliance that would have instantly blinded someone who was observing it from a nearby orbiting planet destined to be burned to a crisp a few minutes later. But the central star-stuff kept getting sucked in the opposite direction—inward. Two suns' worth of material were packed into an ever-smaller volume. Collapsing like a leaking balloon, this perfect sphere became more condensed until its density equaled that of steel.

But the mushrooming gravity at the surface of the shrinking star was far too great to allow any fragment, any atom of this star's body, to stop crumpling. The collapse forced its central material, about 600,000 Earth-mass planets' worth, to continue compacting itself. In just a few seconds, it had shrunk to a ball the size of Earth, and yet it couldn't stop. Indeed, its collapse was accelerating!

An average star like the sun, though gaseous, has roughly the same density as water. But this Taurus supernova's material was now packed so tightly that a sugar cube's volume of it would have outweighed a cement truck.

Smaller and smaller, like Alice. Atoms normally can't get closer to one another than their electron orbitals, because repulsive electrical forces force them to remain separate. But here, the gravity was so intense, all its atoms were squashed enough to make even twentieth-century assumptions about density limits an illusion. On that day in 1054, nature created something science cannot experimentally duplicate.

When the ball of strangeness came to a sudden thudding stop, and an outward-radiating shock wave helped ignite all the star-stuff to set off the titanic explosion we call a type 2 supernova, the star core's atoms' solid particles were in contact. Protons had merged with electrons to become neutrons, and all electrical charges vanished. The sphere, only twelve miles wide, was the only part of that star that remained intact and survived the 1054 supernova. Today, we stare at it in disbelief, knowing it's merely the core of the larger star before it both collapsed and exploded.

A millennium after the supernova, physicists would label such an ultradense object a *neutron star*—the most bizarre, extreme visible object in the known universe. Like all other neutron stars, it could fit between two exits on an interstate. Over half a million Earths' worth of material was packed into this tiny space. As a result, its thickness now matched each of its countless neutrons'. To equal the density of this collapsed star, you'd have to crush a large cruise ship down to the size of the ball of a ballpoint pen while also retaining every ton of the steel.

Imagine a tiny sphere the size of a poppy seed containing all the weight of a luxury ocean liner. That was the density here, except it was not merely a speck of it, but a sphere twelve miles wide. Interestingly enough, this is the density of each atomic nucleus in everyone's body. In a way, the collapsed star core had become one huge neutron one two-hundredth the width of the moon.

Its immense surface gravity warped the very space in the star's vicinity like a fun-house mirror, making background stars shift their positions. The unimaginable gravity forced the tiny dead star's broken atom fragments into a kind of lattice, making its exterior skin a spherical glasslike structure with a hundred thousand trillion times the stiffness of steel. One couldn't damage it or even scratch it, not even with a hydrogen bomb. Nothing in the cosmos could harm it in any way; it wouldn't need to be insured.

A half a mile beneath that impenetrable crust, the star turned liquidy and got even denser, but nonetheless, it boasted a curious super-slipperiness that let the interior effortlessly spin at a different rate than its surface. Despite the extreme crushedness, all friction strangely vanished. And at the very center, only a few miles down, the conditions are a total mystery to physicists to this day. Here our present scientific knowledge comes to an end.

So we'll confine our explorations to its surface, the part we can see, a minuscule star that is the very smallest deep-space object viewable through good backyard telescopes. In one sense, it's a moving target, because any whirling body that shrinks always spins faster, like a ballerina pulling in her arms. So as the Taurus supernova star shrunk, its whirling accelerated. By the time it stopped collapsing and appeared in the sky on that wild day in 1054, it had sped up so much that even today, the entire star maintains a spin of thirty rotations per second.

The newly minted neutron star's magnetic hubs, which leak brilliance at all wavelengths, whirl around with each rotation like a lighthouse beacon. By chance, Earth lies in the right direction to receive a flash of light with every spin.

The problem is, thirty spins per second is too fast for the eye to see, which is why no astronomer knew anything about it until the 1960s. Even through today's best telescopes, the light of the neutron star looks visually steady because nobody can differentiate more than twenty flashes per second, which is the human flicker-fusion threshold. Pulses faster than that appear steady to the human eye.[1]

Once discovered, such tiny, blinking, super-spinning neutron stars were given the catchy name *pulsars*. As others were discovered, astronomers realized that they were all solid, minuscule neutron stars, each a remnant of a supernova, with magnetic poles that sweep past our line of sight with each rotation. With every turn, the pulsar delivers quick bursts of energy at all wavelengths, from radio frequencies to X-rays.

David Helfand, chair of astronomy at Columbia University, showed the author a way to observe the flashes. "I simply wave the fingers of one hand back and forth rapidly in front of the telescope's eyepiece. This crude stroboscope can seem to freeze the blades of a spinning fan and will also reveal this star's pulsations!"

The collapsed star's surrounding magnetism got ramped up bigtime, so it acts as a drag, a brake, and slows the wildly spinning star. The Crab pulsar will whirl just seventeen times a second—with a flickering finally noticeable through amateur telescopes without any hand waving—by the year 4000.

That tiny dense ball, that wildly spinning sanctum of strange-

It was a huge explosion and definitely looks that way even now, a millennium later. At the center of the Crab Nebula sits a double star. The upper-left component is merely an unrelated foreground star. But the fainter member on the lower right is the pulsar itself, the tiny ultradense collapsed remnant of the titanic explosion. This is the famous Crab pulsar. *(Slooh)*

ness, the pulsar has held all our attention. But it's time to look at the bulk of the star, the part we've neglected, the dazzling star material exploding outward that's called the Crab Nebula.

It continued its outrush unabated. Even now, the explosion fragments zoom a distance equal to the breadth of North America every three seconds. It's still an active explosion, as the detritus races ever farther from the collapsed core, and all the while, its brilliance slowly subsides. By the end of 1055, the constellation Taurus had gone back to its original form and looked as if nothing had happened. Except for those written records in the East, the event was utterly gone from human consciousness after a few centuries had passed.

Twenty-two human generations came and went before the invention of the telescope, in 1609. It brought a new level of inspection of the firmament, and still the supernova remained forgotten.

But a century and a half after the telescope's arrival, in 1758, French astronomer Charles Messier looked at that seemingly blank spot in Taurus. On a Paris hotel rooftop, through his six-inch instrument, he spotted the supernova remnant. He saw merely a strange glowing blob.

Messier excitedly assumed he'd discovered a new comet—he'd been searching unsuccessfully for the first predicted return of Halley's comet. He knew that if it brightened enough, it would bring him fame and fortune. But the wind in his sails quickly waned; the blob didn't change position from night to night.

He realized that what had fooled him might fool others, so he decided to catalog all the comet look-alikes that peppered the sky. He made that Taurus splotch the very first item on his list. Though his catalog quickly became wildly famous and ultimately contained one hundred and ten deep-space wonders, the supernova in Taurus was the very first item—Messier number one. This is why, to this day, astronomers around the globe refer to the 1054 supernova as M1.[2]

As telescopes improved and greater detail could be discerned, the supernova's zooming fragments started to *look* like part of an explosion. In 1844, almost a century after Messier, William Parsons, the third Earl of Rosse, used his colossal, unwieldy telescope to look at M1, and he decided that the twisted tendrils resembled the legs of a crab. Lord Rosse named the gas cloud the Crab Nebula, and the name stuck. When astrophotography arrived another half a century later, repeated images of M1 produced a new revelation. Photos taken a few years apart showed it was visibly enlarging at the frenetic rate of nearly one thousand miles per second. Obviously, unlike the frozen majesty of all other nebulae, the Crab has not completed its explosion. The crowds might have gone home, but the fireworks display is not over.

It took the invention of the spectroscope to break down the

Crab's light, and that analysis gave astronomers their biggest shock. Turned out, the nebula's bizarre bluish glow was not starlight, nor reflected light, nor excited gas. It was not the glow of heat or of nuclear energy. The light resembled nothing that had ever been seen on Earth or in the heavens. Much of the glow of this supernova remnant is an exotic phenomenon called synchrotron radiation, a visible emission cast by electrons when they're forced to change direction by superpowerful magnetic fields. The term *synchrotron radiation* was based on its first recognized appearance in a General Electric synchrotron particle accelerator built in 1946, since these, too, accelerate electrons using powerful magnets.

Don't picture the kind of magnetism that wobbles a compass needle. The field near the Crab is a trillion times stronger than Earth's. It would rip that compass right out of your hand. Only that much magnetic energy could blast electrons into violent spiraling geysers so they produce bits of eerie blue light that seem like yelps of protest.

The supernova was more than a big firecracker. It did more than qualify as the second-brightest supernova in earthly skies in all of history, though it missed earning the gold medal thanks to the lesser-observed, far-southerly supernova in 1006 that we reviewed in the previous chapter. And it did more than emit more light than all our galaxy's stars combined. To this day, it is *still* seventy-five thousand times more brilliant than the sun.

In its ephemeral furnace, new elements were forged and then hurled outward to populate the unwalled alleyways of space, to be incorporated into the bodies of new stars, planets, and extraterrestrial creatures. This new material produced every atom heavier than iron, including the uranium we use to make our own firecracker atomic bombs. They exist on Earth solely because of supernovas whose violence unfolded before our planet was born.

The detailed remnant of SN 1054, seen by the Hubble Space Telescope. The debris from the destroyed star still flies away at 930 miles a second. The Crab Nebula has been described as a giant physics laboratory, and it is an enduring monument to the off-the-scale mayhem in our region of the universe one millennium ago. *(NASA)*

CHAPTER 7

TYCHO VERSUS KEPLER: DUELING DETONATIONS

We're almost finished exploring the supernova kind of cataclysm, since not a single brilliant new star has appeared in Earth's sky for the past four centuries. None at all have appeared during the age of the telescope other than faint ones in other galaxies. But before those Milky Way explosions stopped, our quadrant of the galaxy was rocked by two more supernovas—this time over the course of a mere thirty-two years.

The historic importance of these particular blasts was, oddly enough, due to their timing. When the first of these went off, in 1572, Europe was on the cusp of a new era of scientific inquiry that would replace the dark centuries. With an age of enlightenment just beginning, people's take on what was happening produced the first analyses that used the burgeoning scientific method. When the second star exploded, it helped cement this transition to a new enlightened age.

The two blasts are officially listed as SN 1572 and SN 1604, but their popular names are still in use: Tycho's Star and Kepler's Star. The 1572 supernova, Tycho's Star, forces us to become acquainted with a real oddball in the pantheon of science luminaries.

Tycho Brahe was a strange and wealthy Danish noble with a grumpy disposition, a fake brass nose (the result of a college sword duel), and an admirable marriage (he married a minister's daughter, and her commoner status meant that his children lost any claim to nobility). But his life's love was astronomy, and that's why his name is found in every introductory astronomy book.

His schooling required that he study Aristotle's accepted model of the cosmos, which included the "fact" that the sun orbited the Earth. Brahe observed the total solar eclipse of August 20, 1560, at age fifteen, and he was impressed that it had been foreseen but appalled that the prediction had been one day off. He noticed that other astronomical events, like the once-every-twenty-year conjunction of Jupiter and Saturn, were inaccurate too, and he determined to create better data about the heavens.

The telescope's invention still lay a few years in the future, but no matter. With large geared sighting devices with their etched metal notches for degrees and azimuth bearings, observatories like the famed Jaipur Observatory in Rajasthan, India, and England's Royal Greenwich Observatory were making admirably precise notations of celestial positions, and Tycho wanted to do even better.

Thanks to the Tycho family's nobility, the king gave him an island on which to set up an observatory. He proceeded to rule his new little realm and tortured the local farmers with his demands for taxes and manual labor and such, but he also began a lifetime of detailed notes of how the planets moved. He became so skilled that when a brilliant new star appeared in Cassiopeia in 1572, his careful observations showed no parallax for it, proving that it was farther away than the moon and thus disputing the prevailing belief that any and all temporary sky phenomena were mere disturbances in Earth's atmosphere. (One legacy of this long-held belief is the similarity between the words *meteor* and *meteorology*.)[1]

Tycho wrote up his observations in his 1573 book *De nova stella* (*On the New Star*), which became a page-turner among the science lovers of Europe. Actually, *De nova stella* is the title in much later editions; Brahe's original title was *De nova et nullius aevi memoria prius visa stella* ("Concerning the star, new and never before seen in the life or memory of anyone"). His proof that it was not a nearby atmospheric object shook the Aristotelian model that the heavens were immutable. That book on the subject of the Cassiopeia supernova would forever link the Dane with the explosion. In truth, Brahe wasn't flexible enough to fully open his mind to the strange realities of the solar system his patient observational notes were trying to tell him. A movie four centuries in the future, *A Few Good Men*, written by Aaron Sorkin and based on his play of the same name, has Jack Nicholson shouting a line that should have been directed at Tycho Brahe: "You can't handle the truth!" In answer to the long-held Aristotelian and Bible-based Earth-centered views and the rising contrary chorus of Copernican heliocentrists, Tycho offered a muddy, cowardly compromise: Yes, the planets did orbit the sun. But while they were doing so, the sun went around the Earth. It became his life's work to prove this strange thesis, which he'd based partly on the scriptural statement that Earth is "fixed."[2]

But to Brahe's credit, his wrong beliefs about the configuration of the solar system didn't stop him from making careful measurements of planetary positions to an arc-minute accuracy that were at least five times better than anyone else's observations. The king died in 1588 and the new king didn't particularly like Brahe, so he moved to Prague and continued his observations with a new assistant, Johannes Kepler.

Kepler, one of the brightest lights in Europe at the time, was joined by others who pleaded the case for heliocentrism, but Brahe couldn't believe that something as big and heavy as the Earth could possibly be sprightly enough to spin on an axis while also orbiting the sun. But everything changed soon enough. Returning from a banquet one night at age fifty-four, Tycho was unable to urinate, and he soon suffered from severe pain and became delirious. He wrote a deathbed note to Kepler urging him never to use the decades of meticulous observations Brahe had made to support heliocentrism. Before succumbing to the mysterious kidney or bladder ailment (although some in later centuries speculated that he was poisoned by one of his enemies), he wrote his own epitaph in the third person: "He lived like a sage and died like a fool."

Kepler, more brilliant and mathematically advanced than his master, ignored Brahe's deathbed request. He suddenly had access to Brahe's reams of accurate notes on where every planet was situated, and now, unconstrained by Tycho's wacky hybrid celestial model, Kepler asked himself what sort of physical or mathematical orbits would place the planets in precisely the nightly positions they occupied.

What Kepler then created amazes scientists to this day. It made him one of the all-time greats in the realm of celestial breakthroughs. And even more impressive, he did it strictly in his mind, without optical help, since the first use of a telescope to observe the heavens still lay a few years in the future.

But we must now press the Pause button. A mere year before Kepler's astounding breakthrough, another new star had appeared in the heavens. This supernova so soon after Tycho's 1572 star was flabbergasting, and as Kepler was the imperial mathematician to the Holy Roman emperor (having taken over the post upon Tycho's death), his job was to study it. He duly made parallax observations similar to Tycho's three decades earlier; they showed that the supernova lay somewhere in the far distance, beyond the moon and nearer planets. He outlined it all two years later in his 1606 book *De stella nova in pede Serpentarii* (meaning "On the new star in Ophiuchus's foot").

But before we get to the nature of this new star, and of Tycho's, for that matter, we must discuss Kepler's amazing breakthrough. He'd been laboriously trying out various orbital shapes to see which would explain the oddities in the orbits of Mars and the other planets that Tycho's voluminous pages had detailed. Kepler was aware that a few geniuses were saying that the sun's influence got weaker with distance (the concept of gravity still awaited Newton, another lifetime in the future), so each planet's farthest and nearest distance from the sun ought to influence its orbital behavior, and he was prepared to factor in any such fine points if they could easily fit the math and physics of sensible orbital shapes.

Kepler already knew that circular orbits didn't work; they simply wouldn't carry the planets to the observed positions. In 1605, Kepler considered the ellipse. He'd previously dismissed that shape as being a possibility because the idea was so simple, he'd assumed many other astronomers would have thought of it long ago. But when he tried it, it made everything work perfectly. It was simple, really, if the sun occupied one of the two focal points that every ellipse possesses. By the time he'd formulated his three laws of planetary motion, he had precisely explained—and could make exact future predictions for—everything in the solar system. He showed that planets sped up and slowed down as they traveled; Earth, for example, moved fastest when it was closest to the sun in each year, in the first week of January. He demonstrated a simple relationship between a planet's distance from the sun and the length of its year.[3] And it all provided ironclad proof of heliocentrism.

In the four centuries since Kepler's discovery, researchers have called the 1604 supernova Kepler's Star. This is justifiable, given that Johannes wrote the first book about it. But actually, the label was also a way to pay tribute to that truly great man whose immortal three laws of planetary motion were formulated a single year after Earth observers saw that distant sun blow itself to kingdom come.

Kepler endured many tragedies in his life, including the death of three of his children, the witchcraft trial of his mother, and the illness and death of his wife, but he also achieved deserved recognition, especially with the popularity of his textbook *Epitome of*

Copernican Astronomy, which laid out his planetary laws. He saw the increasing embrace of not just heliocentrism but also elliptical orbits in all phases of astronomical motion. A couple of years after his wife died, he married a girl thirty years his junior; they had children and, by all accounts, a happy relationship. His life very much parallels our cataclysm-then-rebirth motif.

Thus the supernova blasts of 1572 and 1604 provided two foci for scientific study at a time when objective truth-seeking, as opposed to explaining events based on superstition or scripture, was gaining ascendency.

Tycho wasn't the one who initially discovered the 1572 star, since it had been cloudy where he was for a week after the star first appeared. Nor did he know that its degree of brilliance had been noted fully five nights earlier. In those pre-mass-communication centuries, no one in one country had an inkling of what people in any other country were observing, which was why he later wrote this:

On the 11th day of November in the evening after sunset, I was contemplating the stars in a clear sky. I noticed that a new and unusual star, surpassing the other stars in brilliancy, was shining almost directly above my head; and since I had, from boyhood, known all the stars of the heavens perfectly, it was quite evident to me that there had never been any star in that place of the sky, even the smallest, to say nothing of a star so conspicuous and bright as this. I was so astonished of this sight that I was not ashamed to doubt the trustworthiness of my own eyes. But when I observed that others, on having the place pointed out to them, could see that there was really a star there, I had no further doubts. A miracle indeed, one that has never been previously seen before our time, in any age since the beginning of the world.

Over the next centuries, astrophysicists kept reexamining that spot in Cassiopeia long after the supernova had faded to nothingness without, apparently, leaving a trace. They found a radio source there in 1952 and, in the 1960s, a faint shell of expanding gas rushing outward at 5,400 miles per second (although more recent studies downgrade the speed to "just" 3,000 miles per second). This is also the source of a violent emission of gamma rays and, especially, X-rays, which continue to stream to Earth's orbiting instruments and which would have been deadly to any and all planets in adjacent star systems. The unfortunate solar system that suffered this cataclysm lies nine thousand light-years away, or at least it did before its destruction.

The event was a type 1 supernova, which, as we've seen, is caused by a white dwarf in a binary-star system gaining material gravitationally from a companion. The small telescopically visible remnant is compatible with such an event, since it is only a type 2 supernova—like the 1054 blast that produced the amazing Crab Nebula—which typically creates giant, eye-catching twisted remnants.

The Milky Way was apparently in a festive mood in the late sixteenth and early seventeenth centuries, so after the blast of 1572, it staged a repeat performance. The supernova of 1604 offered several interesting parallels to its predecessor. As before, the man whose name is permanently linked with it was not its discoverer. In this case, it took a full eight days after the blowup appeared in the sky before Kepler had his initial peek. As before, it was a type 1 supernova in a binary-star system, although this one was twenty thousand light-years away, more than twice the distance of the 1572 blast. That's why Kepler's Star was "only" slightly brighter than Jupiter, at magnitude −3, whereas SN 1572 was magnitude −4 and thus rivaled the dazzle of Venus. Also, as with SN 1572, modern astronomers have detected a small remnant of the explosion.

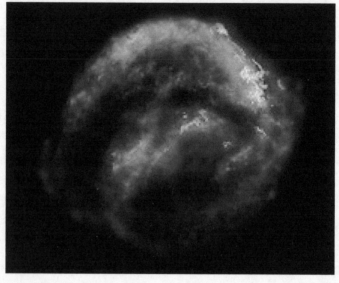

An enhanced image of Kepler's Star, or at least the remnant of it, still wildly expanding and emitting X-rays. *(NASA)*

As far as we know, no other supernovas have gone off in our galaxy since Kepler's Star, and it was certainly the last one that was brightly visible to the naked eye.

Centuries after Kepler, using telescopes, astronomers started to observe supernovas in other galaxies, like the Andromeda supernova of 1885, the first and (so far) only supernova ever seen in our sister galaxy. Just over a century after that, in 1987, another supernova appeared, this one much closer to us, in our irregular dwarf companion galaxy the Large Magellanic Cloud, and it was bright enough to appear faintly to the naked eye to people who lived in the south of Europe and the mainland United States. This author brought a small group to Central America to witness the first naked-eye supernova in four hundred years. It was quite a thrill to see that faint, glowing new "star."

The neighborhood-obliterating explosions that rocked our part of the galaxy nearly half a millennium ago seem to have stopped. Eight known supernovas went off in the Milky Way, half of which lit up the heavens in the eleventh century and then in 1572 and 1604.

But there's been nothing in our galaxy since. When kids first learn about suns exploding, many worry about whether our own sun might someday take the same easy way out. But it can't happen to our own sun, and, more reassuring still, it can't even happen in our cosmic vicinity because no candidate stars are closer than a few hundred light-years. It's more than merely unlikely. It's flat-out impossible.

But other cataclysms certainly can happen, and some involve violence on a far larger scale.

CHAPTER 8

WHEN GALAXIES COLLIDE

L ow-speed, low-mass collisions happen all the time and bother
no one. "Excuse me," you mumble when you brush past someone on a crowded bus. When the mass is ramped up, the damage
starts becoming noticeable, so even a low-speed parking-lot fender
bender makes for a bad day. At what point does an encounter
become a cataclysm? Worlds making contact certainly qualifies, as
we saw in chapter 3. But now let's skip all the preliminaries and
consider the most massive, speediest items in the entire universe.
Normally these bodies mind their own business. But not always.

Of course, collisions may be scary, but they're also fascinating, a
fact to which NASCAR fans can attest. In astronomers' experience
as observers, physical contact between celestial bodies generally
involves only small things, like comets or their fragments smashing
into our atmosphere or hitting an impossible-to-miss target like
Jupiter. We astronomers stared in fascinated disbelief when Comet
Shoemaker-Levy 9 broke into numerous pieces that smashed into
Jupiter one after the other in 1994. Closer to home, casual sky-
gazers observe collisions every night as meteoroids end their ancient
lives with fiery disintegrations through the air overhead.

Star bang-ups would be far more dramatic, but astronomers
have observed only two such events, since each of our galaxy's suns

is typically one million star-widths from its nearest companion. It's no accident that we say they're separated by space.

Galaxies are different. These cities of suns are often so large — one hundred thousand light-years across — that they are usually separated by only twenty galaxy-disk-widths from their neighbors. Hence, stars almost never collide, but galaxies definitely do. Indeed, every galaxy probably undergoes at least one collision in its lifetime. On the largest scales, the universe is a bumper-car ride.

Most galaxies surrounding our Milky Way seem to hover in their own space, threatening no one. Yet everything is in motion, fast motion, and the fact that the Milky Way and the giant Andromeda galaxy approach each other at a speed of seventy miles every second means that an eventual collision is inevitable. The buildup to the impending encounter can be accurately modeled by imagining a twenty-five-foot-wide living room with a football placed on either end. Each football (galaxy) moves its own length every two hundred million years. The collision will unfold sometime sooner than four billion years from now, as we'll see in detail in chapter 32.

And then what? What would this cataclysm look like? Fortunately, we don't have to guess. At a distance from us of forty-five million light-years, NGC 4038, a normal spiral galaxy, and NGC 4039, a barred spiral galaxy like our own, slammed into each other head-on only a few hundred million years ago. The result is a bizarre slow-motion minuet of tidal warping, with gases from the two titans swirling into each other. In wide-field views, enormous tails that look like insect antennae are wildly whipping in intergalactic space, giving this violent merger its name, the Antennae galaxies. These tidal tails were yanked from each galaxy early, perhaps three hundred million years ago, and served as the very first ominous signs of the impending collision.

The smashup process, a self-operating cosmic laboratory that for astrophysicists is almost too good to be true, offers a three-ring circus of simultaneous violence. First we have the cores of the two galaxies, which narrowly missed each other during the initial collision but whose masses are now being swung around like cars on an amusement-park ride. Whipped ever closer by gravity, they'll collide and merge in less than four hundred million years. Computer simulations show that the result will be a huge elliptical galaxy with a single core composed of a jumble of both galaxies' members.

Galaxies in collision. The Antennae galaxies undergo unimaginable violence; their components are hurled at speeds unseen in our own cosmic neighborhood. *(ESA/NASA)*

No future astronomer will likely guess that spirals were ever part of the picture. Those beautiful pinwheel designs, already distorted, will be erased forever. But no stars will actually touch. They're all too far from their neighbors. Merging galaxies merely pass through each other; the invisible fingers of gravity do the sculpting.

But this doesn't mean that all is tranquil. We can see the dusty hydrogen gas, a major spiral-galaxy component, and observe this chaotic mass mixing violently to form vast, dark molecular clouds that block the intense light behind them. Meanwhile, some of the merging hydrogen is condensing into vast assemblies of frenetic new sun births. We're watching millions of stars being created simultaneously, many forming giant blue super–star clusters that emit lethal amounts of sterilizing ultraviolet energy. The gas clouds stamp out factory-line copies of globular clusters, each looking like one of the hundred and fifty spherical star-cities that surround our Milky Way.

This is the very nearest and youngest example of colliding galaxies, and its proximity lets us study its starburst activity in Peeping Tom detail. Many of the Antennae galaxies' faint cobalt blotches are young clusters with tens of thousands of massive baby suns apiece. The pink or red spots popping up chaotically are the campfires of excited hydrogen gas, so-called H II regions, where even more new stars are being hatched, their fierce ultraviolet energy lighting up each gaseous embryonic matrix. Finally, the two giant yellow-orange blobs are the cores of the ill-fated original galaxies NGC 4038 and 4039. They contain the original old stars whose planetary life, if any, must be astonished at the frozen violence that now totally fills their sky from horizon to horizon.

For astronomers, the overall structure is gold. By programming computers to create animations of two interacting spiral galaxies, with separate simulations for varying amounts of invisible dark matter, researchers can determine how much unseen material must be present to fashion the precise shapes they're witnessing. The result? There is four or five times more dark matter than visible material — not ten times more, as many had initially believed.

We are witnessing a movie preview, a trailer, showing us what we humans will not be around to observe for ourselves: what it will look like when our galaxy collides with Andromeda.

If you're skeptical or want to see further proof, you could ponder the Whirlpool galaxy. Here, too, galaxies are colliding, but this time it's an unequal contest. The main member, M51, is one of the most beautiful spirals in the heavens and the only one whose pinwheel shape shows up through large backyard telescopes. It was also the first spiral ever seen. That happened in 1845, when William Parsons, also known as the Earl of Rosse, built the world's largest telescope, an unwieldy giant with a mirror six feet wide.

The first ever observation of a spiral galaxy, published in 1850. The drawing by Lord Rosse also shows the shaft of luminous material connecting it with a much smaller galaxy named NGC 5195 to its right, with which it will merge.

These days, images of the Whirlpool galaxy take one's breath away. They clearly show the first stages of a galaxy collision, analogous to our situation with Andromeda a billion years from now. The initial cataclysmic disruption in the galaxy's symmetry has begun, and it shows itself as a string of material—probably a billion suns and planets—pulled like taffy in an odd line connecting the two island universes.

Two galaxies starting to collide. The main galaxy is the famous Whirlpool, or M51. It's in the earliest stages of merging with the much smaller NGC 5195 above it. *(Matt Francis, Prescott Observatory)*

CHAPTER 9

MAGNETIC VIOLENCE

After about the age of eleven, children grow bored with the magnets that had previously fascinated them. But have they shrugged their shoulders too soon? Sure, it was initially marvelous that a piece of metal should cause another to physically move, correctly suggesting that some mysterious force was reaching out from one to the other. Yet once accepted, the concept became normal; after all, almost no one marvels that every day a ball of fire moves across the sky. But lately, even we grown-ups have become fascinated with magnetism, as it has recently returned as a force with which to be reckoned.

At first, those studying astrophysics found few mentions of magnetism in the textbooks. The sun is the most powerful object within eight light-years of the Mall of America, and its nuclear furnace creates blistering heat and blinding light, but magnetism plays no role in the fusion process that converts its mass to energy. Out there in the cosmos, we learned, there were planetary magnetic fields much stronger than Earth's, but they initially seemed to be like coleslaw: nice but unnecessary. Certainly irrelevant in the energy big leagues. That's how it appeared to astronomers.

Not anymore. During the past few years, researchers have found that neutron stars, already among the all-time-strangest objects, as we saw in chapter 6, can be wrapped in magnetic fields brawny enough to affect the rest of the universe.

The first hint of magnetic trouble came on March 5, 1979, when a burst of gamma rays swept through the solar system at the speed of light. Monitors aboard spacecraft near Venus and Earth suddenly detected radiation that was off-the-scale high.

The deadly torrent lasted for only a fifth of a second, but in that eyeblink, some mysterious object had given off twice the energy the sun had emitted since the Great Pyramids were built. The burst could not be explained by any known phenomenon. Eventually the culprit was determined to be an invisible neutron star in a neighbor galaxy.

All that power from a tiny neutron star, a dead ex-sun that didn't even boast a nuclear generator? Maybe astronomers needed to take another look at these stars.

As previously discussed, all neutron stars are tiny. They also boast solid surfaces—indestructible, half-mile-thick crusts floating atop a bizarre fluid of subatomic particles. This material is hyper-packed; an apple-seed-size speck outweighs a freight train loaded with iron ore.

As another reminder, a neutron star forms when a massive star collapses and sends supernova brilliance shooting outward and a tiny remnant core imploding inward. That core—now a twelve-mile-wide sun—spins crazily, often hundreds of times a second. Such frenzied motion causes its magnetic field to wrap around itself, intensifying the field lines. What's new about all this are indications that for a brief period in a neutron star's youth, its magnetic field can reach a strength of a thousand trillion gauss. By comparison, Earth's field is less than one gauss. With fields a million billion times greater than a refrigerator magnet's, such stars are catchily called magnetars.

More bursts arrived after that first eye-opener. The biggest came on December 27, 2004, and that time the gamma rays were strong enough to ionize atoms in Earth's own atmosphere. It produced a glow bright enough to be reported by casual stargazers—the first and so far only deep-space object to make something light up on our planet, at least in modern times. Then, in 2007, another burst swept past Earth, this one coming from a different direction.

A dozen magnetars have now been discovered, but only one, in Cassiopeia, can be seen through telescopes. The others were detected solely by the radiation they emitted. But over a million other unknown magnetars are thought to loiter unseen in the dusty hallways of the Milky Way.

I spoke with Vicky Kaspi of McGill University, who won the 2004 Herzberg medal for her studies of these fantastic objects. She's obsessed with them.

"Magnetars are the only deep-space objects to directly affect Earth," she said. "The 2004 burst changed our ionosphere from night to day. Some fishermen in the Arctic saw a sudden aurora at that moment."

And get this—magnetars are powered not by nuclear energy, like the sun, nor by the loss of kinetic energy from rotation slowing. "In magnetars we see a unique source of power: *all* of its energy comes from the gradual loss of its magnetic field," Kaspi explained.

The intense magnetism bends and deforms a magnetar's solid crust to produce starquakes. These are nothing like the tremors we get here, earthquakes that can merely destroy a city. No, a neutron star's ultradense starquakes release titanic bursts of energy that actually create antimatter! When these electrons and antimatter positrons combine and annihilate each other, they convert 100 percent of their mass to energy, producing the lethal gamma rays that sweep through the universe.

Meanwhile, the magnetism slows the star. In a mere ten thousand years, according to current thinking, the magnetic field weakens to a paltry two trillion times greater than Earth's. Then the starquakes stop and the gamma rays die out.

In short, magnetars embody the physics of extremes, extremes of density, gravity, and magnetism. No wonder they're so much fun. As long as we keep our distance.

Imagine a future astronaut approaching one. When she's as close to it as the moon is to Earth, the magnetar is still too small to appear as anything but a shapeless point. When she's half that distance from it, all of her credit cards mysteriously get erased. She now can't buy a souvenir in the magnetar gift shop, but curiosity drives her forward. When she's twenty thousand miles from the magnetar, it still appears as just a dot, but its magnetism now pulls every atom in the astronaut's body into long, strange needle formations.

If her vital signs are being monitored, her health-insurance company will start to get fidgety and its employees will begin thinking of excuses to cancel her policy.

If she's wise, she'll now do a U-turn; otherwise this will be a fatal attraction. And although these magnetic consequences are accurate and not speculative science, they're offered to show that, scattered throughout our galaxy, there are bizarre pockets of off-the-scale magnetism that are anything but harmless.

In our own solar system, all the places of strong magnetism are, at the moment, safely isolated from us. Our own planet's magneto-sphere is more than nonharmful; it's protective and blocks most cosmic rays from reaching the surface. But if and when humans venture outward to Jupiter's strange and very possibly life-supporting moons, they will have to figure out a way to deal with that planet's deadly magnetic field, which traps and accelerates subatomic particles to create a lethal radiation hazard.

We already dream about venturing to Europa, the satellite with warm saltwater oceans that might be harboring life. But standing on its ice-covered surface means being zapped by 540 rem, or 5,400 milli-sieverts, of radiation per hour, which would be fatal in about ninety minutes. The hazard arises because of Jupiter's superstrong magnetism.

And later we'll see why some worry about future Earth pole reversals, which could remove the planet's magnetic shielding and expose the entire world to increased deep-space radiation. Thus, magnetism is no problem for us this year, but it cannot be ruled out as a cause of future mayhem.

CHAPTER 10

THE LETHAL ANTIMATTER FOUNTAIN

These days, the eccentric, brilliant physicist Paul Dirac is remembered mostly by science nerds. Yet it was he who made the amazing prediction in 1928 that there should exist a substance we now call antimatter. When it was actually discovered seven years later, Dirac should have won the Nobel Prize and become a household name like Einstein. Well, he did indeed win the Nobel Prize in Physics. But if you mention his name to your friends, you'll probably get only blank stares.

Antimatter looks and behaves just like ordinary matter. Observing it, you couldn't tell antimatter from antipasto. An antimatter star would look just like a normal one, and no amount of high-tech analysis such as spectroscopy could help you tell them apart. But if an antimatter object touches anything made of conventional material, both vanish in a violent flash.

Unlike most of the exotic particles and objects that live in the weird section of the modern cosmological zoo, antimatter is simple to comprehend. It's merely ordinary matter with all electrical charges reversed. An antiatom's nucleus is negative instead of positive. And it's orbited by positive instead of negative electrons, particles that are therefore called positrons.

Just because they're logically simple doesn't rob them of mystery — or danger. Every version of the Big Bang theory says that equal amounts of matter and antimatter should have been created 13.8 billion years ago, yet somehow we find ourselves in a matter-dominated universe. What happened to each potential bit of anti-planet, antiocean, and antifreeze? Currently, the best explanation is that, contrary to the theory that says nature has no preference for one over the other, there is a tiny bias in particle-antiparticle events, and matter seems to be slightly favored.

And a good thing too. A universe with much antimatter would be a dangerous place. No explosion is more powerful than when matter and antimatter meet. It's a 100 percent $E=mc^2$ conversion of the masses of both objects. If a one-gram pencil eraser touched an anti-eraser, it would release two trillion trillion ergs of energy, enough to light every bulb in the United States for ten days. Ounce for ounce, antimatter is 143 times more energy-potent than the sun's fusion furnace or an exploding H-bomb.

So, yes, the most dramatic thing about antimatter is what happens when it encounters ordinary materials. Every cigarette butt on the street is a cauldron of super-energy trapped in a dormant (and disgusting) form. Release it and you could power your house for a thousand years.

Nuclear fusion is how the sun shines, but it's a wasteful process. H-bombs and the solar core unleash energy with only a 0.007 (double-oh-seven) efficiency, meaning just seven-tenths of 1 percent. But when matter and antimatter meet, their masses convert to energy with 100 percent efficiency, so pound for pound, anti-matter fuel is twelve dozen times more powerful than thermo-nuclear weapons, and it would work with anything at all — junk mail, corpses; any form of matter is as good as any other. Just let it come into contact with antimatter.

But where to get the antimatter? Nobody's selling any, not even on eBay. There's no generic version. It's made in particle accelerators like Fermilab, but it's not cheap. Current worldwide production is about ten nanograms (ten billionths of a gram) per year. The biggest chunks of it are whole antiatoms, which were first created in 1995. So, yes, scientists long ago learned how to create antimatter positrons, which have moved from the sci-fi world to the high-tech marketplace. They can produce exquisite images of the body; positrons are the *P* in medical PET scans.

Antimatter is hard to keep too. It can't be stored in any jar because it explodes when it touches anything and would annihilate the container's wall. Like leftover Chinese food, antimatter should be kept in the fridge, but way down on the bottom shelf in the absolute-zero bin, where there's barely any motion. (Heat is just atoms in motion. A very cold temperature is the same thing as saying that atoms are lethargic.)

If ionized so that it has an electrical charge, the antimatter will allow itself to be suspended in a magnetic field. A moving field, like those used for levitating trains, could guide such fuel to a combustion chamber to meet bits of regular matter, even stuff from the vehicle's toilets. Yes, you could go far and fast with an antimatter engine under the hood.

That's what the starship *Enterprise* uses for propulsion. Yet even there in the twenty-third century—as Scotty kept reminding us when he yelled in his brogue, "Captain, I'm losing containment!"—there are perils in the system.

Neither Kirk nor Picard liked hearing that little anxiety-inducing "losing containment" comment from his chief engineer. Scotty rarely spelled out the problem, but everyone on the bridge knew that if the magnetic vacuum failed for any reason more serious than a tripped circuit breaker, the walls of the ship would meet the antimatter and result in a detonation that would be far more energetic than any of their weapons.

Astronomers' job at the moment is to see whether this is actually happening and, if so, whether it is indeed creating unusually weird cataclysms in parts of our galaxy. Although a peek at the night sky can't tell you whether a particular star is made of matter or antimatter, there are good reasons to believe that ours is a matter neighborhood in a matter galaxy. The explosive contact between matter and antimatter produces gamma rays with a distinctive energy signature of 511,000 electron volts. Thus, if any antimatter fringe material contacted ordinary particles, they'd produce unique energy halos, and these had not been observed.

At least, not until 1997. That's when the Compton Gamma Ray Observatory discovered positronic geysers right here in our own Milky Way. This frenzied energy from the galaxy's center extends out thirty-five hundred light-years. The violent gamma-ray emissions are definitely caused by matter-antimatter collisions. They would be lethal for all the worlds in that vicinity and all planets orbiting them. It's not small-scale mayhem.

But what could be the source of these positrons?

Whatever it is, it's doing an efficient job, because at the source, ten million trillion trillion trillion positrons are encountering matter and being converted to energy each second. In the Milky Way's central bulge, about two hundred billion tons of positrons were annihilated in the time it took you to read this sentence. In that interval, some five hundred trillion tons of matter were involved in the process, and this happens day in and day out. Where is this antimatter factory? There is no source of activity visible to our telescopes or detectors. Since positrons in the vacuum of space live for millions of years, they might be coming from an old supernova or lots of them, maybe from a long-ago era when the Milky Way's bulge was continually exploding like fireworks. Or perhaps the antimatter originates from some kind of black-hole antimatter production line. Other usual suspects that have been considered and eliminated include pulsars, quasars, and cannibalized satellite galaxies. In each case, there seems to be no way for so many positrons to get created, let alone be transported so far from their birthplace.

The problem is not merely one of genesis but how fifteen billion tons of positrons per second are being hurled like water spray thousands of light-years above the galactic plane. Theorists' most intriguing idea (some would say more like a wishful hope) is that dark matter, whose nature is still mysterious, might be generating the antimatter through some as yet unknown interaction.

Translation: No one has a clue as to the source. But that doesn't mean it's not there. It's an ongoing cataclysm, a keep-away police line that will rope off the innermost regions of our Milky Way for a long time to come.

CHAPTER 11

DANGEROUS BUBBLES

We are all bathed in a wide variety of electromagnetic radiation (a fancy way of saying *light*). We are surrounded by visible objects of various colors, as well as by sections of the electromagnetic spectrum our eyes never see. We can feel the infrared from the sun because it makes our skin's atoms jiggle faster, which gives us the sensation we know as heat. But we cannot feel the radio waves that are just as omnipresent in our lives. Nor can we sense X-rays or microwaves or gamma rays, even though they, too, crisscross the cosmos.

These various forms of light differ only in how far apart their waves are spaced. Radio waves are miles apart from crest to crest, so when they bump into things, they jostle them only gently. Microwaves, however, possess just the correct spacing to make water molecules jiggle, which is how a microwave oven works. Causing water molecules to move faster is simply another way of saying "heating them up," and since water is found in virtually all foods, it's a quick and easy way to turn frozen pizza into something desirable.

Waves of greater sizes than visible light are gentle; they cannot tear apart atoms (a process called *ionization*) and therefore cannot harm genes and chromosomes, so they cannot (we think) cause cancer. But close-together waves are a different story.

Beyond the violet end of the spectrum lies ultraviolet. These rays can indeed induce genetic damage. Australians with their sunny climate have a higher skin-cancer rate than Americans simply because a 1 percent rise in lifetime UV exposure produces a 1 percent increase in cancer risk.

Waves that are closer together than UV are far more harmful. These are the X-rays and, especially, gamma rays, which pass right through animal bodies, damaging chromosomes along the way. Happily, these forms of electromagnetic radiation are extremely rare on Earth. Though the sun creates copious gamma rays in its core, the bulk of its body acts like a kind of insulator, causing them to be absorbed and then emitted as softer kinds of light; what ultimately escapes from the visible solar surface is a fifty-fifty blend of infrared and visible light, neither of which is harmful.

Though the sun doesn't let many gamma rays leak out, they are indeed created here and there in the universe, always by extreme violence like supernova explosions, but our atmosphere blocks them. Hence, the only way to study gamma rays is to send up a special orbiting spacecraft specially designed to map the brutal, mysterious gamma-ray nether regions. That is why, in 2008, NASA launched its Fermi Gamma-Ray Space Telescope.

At first, the probe detected individual exploding stars in distant empires, separate little gamma-ray pinpoint hotspots. But then it found something else—something that was more than merely peculiar. It was something that made no sense whatsoever.

In November of 2010, astronomers using Fermi made an astonishing announcement. Emanating from the center of our Milky Way galaxy were two bubbles made solely of lethal gamma rays.

This would have been strange enough if the bubbles, expanding at 2.2 million miles an hour, were concentric—a bubble within a bubble—with both centered at the galaxy's core. But no. The two enormous spheres each hover in seemingly empty space above and below the black hole in the Milky Way's nucleus. The surfaces of these bubbles just barely touch at the galactic center to form a squat hourglass shape. The entire structure looks like the number 8 or—to be deeper and more intriguing—an infinity sign twisted ninety degrees.

Since gamma rays are the bad boys of the electromagnetic spectrum—the highest-energy photons in the universe—they do not reliably reflect off objects the way visible light does. In fact, they are penetrating. Stars do not normally emit them, so why should there be a dense gamma-ray swarm at our galaxy's center, where nothing whatsoever is exploding? It's the unmistakable sign of extreme violence. It's proof of a cataclysmic event, probably in the not-too-distant past. And yet, these days, the Milky Way's core is about as energetic as a steamy July Memphis lunchtime.

The bubbles are sharp-edged and well defined and nothing short of enormous. The top and bottom of the figure eight extends from twenty-five thousand light-years north of the flat galactic plane, the pancake-like structure of our galaxy, to the same distance beneath it. From Earth's sideways viewpoint twenty-five thousand light-years from the center, the hourglass stands a whopping forty-five degrees above and below the galactic core in Sagittarius. It takes up half of our southern sky.

Theorists need to explain more than just what could have produced this kind of extreme energy, which is equivalent to a hundred thousand exploding supernovas. They must also explain the off-center nature of the cataclysmic bubbles, since each seemingly surrounds nothingness.

Jon Morse, director of astrophysics at NASA headquarters, summed up the discovery at a press conference: "It shows, once again, that the universe is full of surprises." This gargantuan hourglass, which is often referred to as the Fermi Bubbles in honor of the orbiting gamma-ray telescope that found them, is now regarded as an *entirely new type of astronomical object*.

In trying to come up with some explanation for our galaxy blowing bubbles at temperatures of seven million degrees Fahrenheit, many astrophysicists express a gratifying unanimity; they say, "We have no idea." Others, starting perforce from square one, have posited a couple of vague possible causes. The first theory is that, perhaps a few million years ago, a burst of star formation at the galactic center created numerous massive stars, all with high-speed winds of escaping high-energy particles. Since this alone could not begin to explain the superhigh power within the bubbles, that theory further imagines that many of these stars blew up into supernovas simultaneously.

If such ultra-mayhem happening simultaneously sounds too implausible—which indeed it does—let's go to possible explanation number two, which is that the massive black hole at our galaxy's center, which has the same weight as four million suns, had a brief feeding frenzy when enough captured star material was accelerated inward by its awesome gravity. Then, perhaps, that black hole could have developed something it does not presently have: twin jets of outrushing material. We see such blue jets exploding from the supermassive black holes in a few other galaxies. These

The dusty horizontal line is our own Milky Way galaxy viewed from an edgewise perspective. The two enormous and violent gamma-ray bubbles barely touch at the black hole in our galaxy's core. Their origin is a mystery. *(NASA)*

jets could have possibly deposited energetic material above and below our galaxy's plane, although why bubbles then emanated from those positions is anyone's guess. In short, what we see is just too violent, too deadly, too inexplicable, and too impossible to have been an aftereffect of some occurrence. Rather, it has the lethal ambience of an event of unimaginable violence, a super-cataclysm of unknown etiology.

The ultimate answer, though it must involve ultraviolence, could be even stranger. Might these outrushing shells of H-bomb-level energy be the long-sought signs of dark matter—that mysterious substance now known to make up one-quarter of the universe? Could dark matter be meeting its opposite entity (whatever *that* is) in total annihilation, the way matter and antimatter do?

More likely, however, this is something else entirely, some new phenom that will actually get in the way of the dark-matter hunt. As Fermi research team leader Douglas Finkbeiner put it to the author, "This just confuses *everything*."

So we observers twenty-five thousand light-years away can only gawk and wonder about what kind of unimaginable cataclysm created this most violent electromagnetic energy in the known universe. We can't visualize anything this powerful. We don't know where the explosion was located, even though the explosive remnants now occupy fully half of our southern sky. All we detect through our orbiting instruments is the angry swarm of detritus streaming outward in two enormous spheres too energetic for our eyes to behold. Our organ of vision simply was not designed to perceive such photonic brutality.

CHAPTER 12

THE EXPLODING GALAXY NEXT DOOR

For the average stargazer, the first recognizable star pattern in childhood is usually the Big Dipper, which always hovers in the north and is overhead every spring. Those who can get away from city lights see the Big Dipper as isolated, since its location far from the creamy glow of the Milky Way helps keep it segregated from any bright stars. It's the last place you'd expect to find violence.

For astrophysicists, the Big Dipper's background darkness is exactly the quality that beckons, since here we look through the thinnest amount of foreground Milky Way material to get an unobstructed view into intergalactic space. It's where a Hubble astronomer chose to take the first famous Hubble Deep Field photograph by letting that orbiting observatory stare for a patient one hundred hours at a spot of sky the size of a sand grain, all the while collecting light into a single image.

And it was in this same area, on the final day of 1774, that German astronomer Johann Bode found two pale "nebulous patches" less than one degree apart.

Five years later, French comet hunter Charles Messier's friend and assistant Pierre Méchain rediscovered the smudges, which were quickly added to Messier's catalog of comet look-alikes as numbers 81 and 82. Telescopes improved in the nineteenth century,

The Hubble Deep Field photograph was a hundred-hour exposure of a tiny section of sky beyond the Big Dipper. It shows more distant galaxies than foreground stars. *(NASA)*

and Messier 81 was seen to be a beautiful if ordinary spiral galaxy. But its companion M82 remained a puzzle. Even today, seen through large amateur telescopes, it resembles neither a spiral nor an elliptical galaxy; it looks rather like a cigar crossed by dark dust lanes. Astronomers eventually shrugged and listed it as "irregular," a status it maintained until the great twentieth-century telescopes revealed unique explosive-looking details that demanded some sort of explanation.

Its status duly shifted, and it was often listed as an "exploding galaxy," a description still common in textbooks. But what did this mean, exactly? Galaxies are huge assemblages of stars, dust, and gas. Not a single star was seen to be exploding in M82. No supernovas were observed there, although astronomers detected radio evidence of a sole hidden one in 2007. Gas doesn't explode, and neither does dust. So although M82 certainly looked like someone set off dynamite in its core, they decided there was no reason to call it an exploding galaxy. At least, there wasn't until 1970. That's when astronomers discovered that an unprecedentedly enormous explosion had detonated in that galaxy's core one and a half million years ago (which in astronomy terms is like the day before yesterday). They found debris hurtling outward at a speed of six thousand miles a second. That means the material travels twice the coast-to-coast width of North America each and every second. That's 3 percent of the speed of light, which wouldn't sound impressive if the detritus were merely subatomic electrons and such. But here, visible macroscopic solid chunks, like shell fragments, can be seen flying outward. So another revision was required—the label *exploding galaxy* was reinstated. And yet, this wasn't the source of the blast-like deformation that made M82 look like no other galaxy.

So what was going on here? M82's long cigar shape resembles an edgewise spiral galaxy, and better photographic techniques started revealing fountains of material thrown violently upward and downward from a flat central plane. At the same time, astronomers measuring its distance from us found that it and its normal-looking neighbor M81 were just twelve million light-years away. Among our nearest galactic companions, it was sufficiently close by for detailed study, and they believed that the origins of these violent-looking features would be determined soon enough.

Radio telescopes provided a clue in 1953, when astrophysicists found that M82 emitted intense radio energy. The galaxy was then given yet another label; now it was Ursa Major A, indicating that it was the brightest radio source in the Big Bear constellation. Invisible sections of the electromagnetic spectrum promised to yield more information, and in the 1990s, the orbiting Chandra telescope found copious X-ray emissions. These emanated six hundred light-years from M82's center and fluctuated strangely. They appeared to come from the first-ever intermediate-mass black hole, weighing somewhere between two hundred and five thousand suns. This unique object, however, was definitely not responsible for M82's bizarre appearance.

Where M82 really dazzles is in yet another part of the spectrum — the infrared, commonly regarded as heat. *In infrared light, M82 is the brightest thing in the universe.* Moreover, infrared images reveal two buried spiral arms, proving that M82 is indeed an edge-on spiral, albeit one that has undergone some strange cataclysmic metamorphosis.

This infrared or heat energy provided a major clue. We see an infrared excess in only a few other galaxies, most notably Centaurus A and NGC 5195, the companion to the famous Whirlpool galaxy. These galaxies have one thing in common: They are each colliding with another galaxy. They are the locations of current cataclysms.

Suddenly, everything made sense. Looking just to the left of the Big Dipper's bowl on April evenings, we are seeing nothing less than the nearest galactic collision. Turns out, M82's companion M81 is not irrelevant, like strangers in the same elevator. The two are in a gravitational minuet, whirling around and through each other, a process that will ultimately merge them into a single entity. Even now, the two are separated by only about a hundred and thirty thousand light-years, little more than the width of our Milky Way. A few hundred million years ago, their last close encounter stirred up M82 and roiled its gases to create wild bursts of new star formation.

The innermost thousand light-years of M82 is a prodigious nursery, where stars are being born ten times faster than they are within our own Milky Way. Here, the prolific production of billions of new suns and the turbulent, swirling flow of gases is indeed "explosive," and that's what incites all that infrared and radio noise. Even in visible light, M82 shines five times more brilliantly than our own galaxy, some of that light coming from a hundred newly minted globular clusters found by the Hubble telescope.

Whether labeled a starburst galaxy or an exploding galaxy, the nearby M82 emits more heat (or infrared energy) than any other object in the universe. *(NASA)*

It was clearly time to reclassify M82 once again, a seemingly annual ritual. It is now labeled, hopefully permanently, a *starburst galaxy*. Although its days as a separate entity are numbered, thanks to its imminent merger, M82 is nonetheless enjoying a final heyday, an ironic springtime celebrated exuberantly by youthful new suns and dazzling bright lights.

In its sleepier regions, this galaxy in our neighborhood is benign. If you lived there, you'd never know that life-forms twelve million light-years away called it an exploding galaxy. But in the galaxy's wildly disrupted sections, all hell is breaking loose. There, supernovas are poised to go off like firecrackers and wipe out entire planetary systems. It's the apocalypse on steroids.

CHAPTER 13

WAITING FOR A CARRINGTON CATACLYSM

Our solar system's neighborhood bully is, of course, the sun. Even in the nineteenth century, astronomers knew that it ruled the solar system and that if there was any true peril to our planet, it would come from that brightest object in the sky. And yet no one fully appreciated its temper.

People knew that it was powerful and even that its emissions varied in an eleven-year cycle, but no one knew that it sometimes threw hissy fits that could greatly affect the planet.

This is where Richard Carrington enters our story. He was a British astronomer obsessed with the sun. Back then, in the mid-nineteenth century, astronomers were mostly focused on studying the sun's strange dark storms, and Carrington observed them every clear day (which in England meant very rarely). Still, he was destined to witness a violent phenomenon nobody had ever seen before. What he observed in 1859 was so strange and violent, it went down as a historic event that makes solar researchers nervous even now, a hundred and sixty years later.

Carrington was self-educated. He had no college degree, nor did he teach at any school. His Wikipedia biography cites him as an amateur astronomer, but the truth is that in that era, many of the greatest discoveries were unearthed by nonprofessionals. And in fact, by midcentury Carrington had already made breakthrough solar discoveries that had earned him the Copley Gold Medal — that era's equivalent of the Nobel Prize. So we shouldn't regard him as a mere tinkerer or gentleman hobbyist. His influence was sufficiently great that today, solar experts allude to the current solar sunspot cycle, number 24, as the newest member of "the Carrington cycle."

The seminal event in Carrington's life arrived on September 1, 1859, when he peered at the sun through a protective filter on his modest four-inch refractor telescope. Unusually intense storms had been parading from left to right across the sun's disk all week, and he took careful notes of the day's activity. Suddenly, at 11:18 a.m., he watched in disbelief as a large sunspot group near the top of the sun grew more and more dazzling.

Despite the telescope's protective filter, Carrington could barely look at it. At the same moment, a hundred miles away, Richard Hodgson, another amateur astronomer, was peering through a six-inch telescope, and he later confirmed that he too had seen "a very brilliant light…most dazzling…much brighter than the sun's surface."

Brighter than the sun? And yet this odd story was just getting started. Simultaneously, instruments at the Kew Observatory recorded a disturbance in the Earth's magnetic field. Could this finally be the long-sought-after evidence that events on the sun did actually affect our planet? The answer arrived that very night. Brilliant auroras lit up the skies around the globe, even over southern regions like the Caribbean.

These northern (and southern) lights exhibited more colors than their usual pale green. They contained vivid deep reds, alarming millions of people who had never seen or imagined that the heavens could contort in such brilliant, twisted explosions. They were so bright over the western United States that people got up and made breakfast, assuming dawn had arrived. In New York City, gawkers crowded on rooftops and sidewalks.

We now know that Carrington had witnessed a solar flare. Such ultrapowerful explosions have the energy of a million hydrogen bombs and hurl tons of material into space at high speeds. If a big flare or an even greater accompanying cataclysm called a coronal mass ejection (CME) happens in the middle of the sun or, especially, on its right or western side, the solar detritus will follow the sun's spiraling magnetic field on an easy path all the way to Earth.

While Richard Carrington watched in amazement, this one flare got so luminous it doubled the brightness of the solar disk he was observing. He could not, of course, see the superpowerful X-rays that were explosively streaming from that same spot. (If X-rays were visible to humans, their brilliance would dwarf the visual emissions.) And he couldn't know how the explosion was affecting the world.

When that incredibly high-speed piece of the sun's energy hit Earth on the night of September 1, the shotgun blast created the most powerful geomagnetic storm ever witnessed. Back then, before satellites and electronics or even power lines and pipelines existed, the only wires capable of carrying current were the telegraph lines that had been erected all over the United States and Europe just twenty years earlier. These started sizzling and popping in showers of sparks. The impossibly high-energy current instantly found its way into populated areas. Sparking equipment alarmed countless telegraph offices across Europe and the United States as frightened operators leaped from their seats. Some did not move in time and were found unconscious on the floor.

After the Carrington event, as it started to be called—an event that lasted only five minutes—it was a full decade before any astronomer saw another flare.

The second-worst geomagnetic storm in recorded history occurred half a century later, from May 13 to May 15, 1921. According to a series of articles in the *New York Times* that month, pieces with headlines like "Sunspot Aurora Paralyzes Wires," the entire signal and switching mechanisms of the New York Central Railroad were knocked out of operation, and a fire raged in the control tower at Fifty-Seventh Street and Park Avenue; a fire in the Central New England Railroad station destroyed the whole building. At the same time, telegraph operations throughout the country were disrupted.

Again, this was before the advent of electronics, pipelines, satellites, GPS, jet aircraft flying polar routes, a manned space station, and long-distance high-voltage power lines. People wondered what a Carrington storm would do in modern times.

We found out on March 13, 1989. The solar violence was less brutal than it was during the 1859 and 1921 events, but it did the job. At 2:44 a.m. on March 13, sun-induced surges started creating havoc in Quebec's electrical-power grid. The hundred-ton capacitor number 12 at the Chibougamau substation tripped and went offline. Two seconds later a second capacitor blew, and then a hundred miles away at the Albanel and Nemiskau stations four more capacitors went offline. When yet another went offline, and five transmission lines from James Bay tripped, the entire 9,460-megawatt output from Hydro-Quebec's La Grande Hydroelectric Complex was cut off. Within a minute the Quebec power grid had collapsed. The province of Quebec was blacked out. Three million people were in darkness. Over half a million of these depended on electricity for heat, but nothing could be done. Dawn broke and the workweek began with the Montreal metro silent and useless. The city's main airport, Dorval, stripped of its radars, also closed.

The storm came within a hairsbreadth of engulfing the big country to the south. United States electrical grids were perilously close to a series of shutdowns; New York Power had lost 150 megawatts the moment the Quebec power grid went down, while the New England Power Pool lost ten times that amount. Ninety-six New England utilities lost power. A large step-up transformer failed at the Salem Nuclear Power Plant in New Jersey—one of about two hundred separate events in the United States that included generators tripping out of service and voltage swings at major substations. And yet, the United States managed—just barely—to avoid a series of cascading blackouts.

Meanwhile, high above us, several satellites were damaged. Some had their semiconductors fried by the intense solar bombardment, and they tumbled out of control for hours before they could be brought back. The space shuttle *Discovery*, launched a few days earlier, suddenly started having problems too; a sensor on one of the critical tanks supplying hydrogen to a fuel cell showed anomalous pressure. When the solar storm ended, the mysterious readings vanished.

The next solar attack came fourteen years later, just before Halloween of 2003. NOAA, the National Oceanic and Atmospheric Administration, reported that it was the second-fastest recorded journey of solar material to Earth, twice as speedy as the normal three-to four-day travel time. The damage it caused was ultimately costly, and NOAA's Space Weather center issued a detailed, sobering analysis. The storm's wide-ranging effects included:

"Strong geomagnetically induced currents (GIC) over Northern Europe caused transformer problems and a system failure and subsequent blackout."

"Radiation storm levels high enough to prompt NASA officials to issue a flight directive to the ISS astronauts to take precautionary shelter."

"Airlines took unprecedented actions in their high latitude routes to avoid the high radiation levels and

communication blackout areas. Rerouted flights cost
airlines $10,000 to $100,000 per flight."

"Numerous anomalies were reported by deep space missions
and by satellites at all orbits. GSFC Space Science Mission
Operations Team indicated that approximately 59 percent
of the Earth and Space science missions were impacted. The
storms...caused the loss of the $640 million ADEOS-2
spacecraft [which carried] the $150 million NASA
SeaWinds instrument."

Yet this was only a moderate solar burp compared with the 1859
Carrington maelstrom. This new Halloween Storm, as space-
weather experts were calling it, didn't even happen at solar maximum, when sunspots and other solar violence hit their peak. What
would a bona fide 1859-level event do to us in today's high-tech
environment? In May 2008, a U.S. government workshop was con-
vened to answer that very question. For the first time, an all-star
team of space-weather experts was assembled under one roof. Their
conclusions were anything but reassuring.

They estimated that a Carrington-level sun storm would produce damage of one to two trillion dollars during the first year alone
and that this "severe geomagnetic storm scenario" would require a
recovery time of four to ten years. Were they exaggerating? Why
didn't the public know about a threat of such a grave nature?

As one presenter in the conference noted with alarm, "Space
weather events are a classical example of what social scientists call a
low-frequency/high-consequence event." That is, something that
has the potential to have a significant social impact but that does
not occur with the frequency or discernible regularity that forces
society to develop plans for coping.

A very strong solar storm creates high-energy electromagnetic waves, including extreme ultraviolet and X-rays, that come in at light speed. They ionize our atmosphere down to forty-five miles and can produce radio blackouts. This is why the Earth's magnetic shield was under attack immediately in 1859, even while Carrington was still gawking at the flare.

The solar violence then unleashes a second wave of assault consisting of a radiation storm. The level can approach a hundred million electron volts, which can be a danger to airline passengers and extremely hazardous or even lethal to astronauts. This radiation can even make it all the way to the ground.

The third and final invasion from a solar explosion is the geomagnetic storm. This is what causes electric grid collapses and power blackouts. Such a storm also sends currents of hundreds of amps along pipelines; it damages transformers, destroys radio communications, knocks out satellites, and downgrades the GPS system.

Our defense against all this? Currently, the first line is an armada of specialized satellites. The most famous is probably GOES, which stares at the sun and the Earth simultaneously from Earth orbit and measures the effects of solar storms once they've arrived. It looks at the sun in the light of X-rays and has a detector to count protons and other incoming bullets.

SOHO, which is parked nearly a million miles sunward of us, stares at the sun continuously and can see solar flares and CMEs with its coronagraph. It watches them being created and thus usually provides us with two to four days' warning.

ACE is possibly the most valuable of all, even if it has exceeded its shelf life and doesn't yet have a replacement. ACE is parked at the Lagrangian L1 point, where the sun's gravity balances our own, and it alone is able to measure the density and magnetic polarity of the solar wind that sweeps past the spacecraft on its way to Earth. Hence, ACE gives us *specific* warnings one or two hours prior to a space-weather event that will almost surely affect us (solar detritus affects us only if it has its own magnetic field that is oriented opposite to Earth's). Even cooler for science nerds, ACE's real-time information is available to the public, meaning that you can see the same thing space-weather experts observe. To get the current readouts (which look like EKG squiggles) of the solar wind's magnetic polarity, speed, density, and more, just go to http://www.swpc.noaa.gov /ace/MAG_SWEPAM_24h.html and be your own sun observer.

Then there's STEREO, a pair of machines that observe flares and CMEs to perfection and show dark coronal holes where the fastest solar wind streams unimpeded into space to arrive here a few days later. If a CME is being blasted straight toward us, it shows up as a symmetrical explosion; it *looks* like it's coming at us. Stereo also gets to peer around the sun's back side, so it can see storms that are about to rotate into the hemisphere facing us.

A newer satellite, launched in May 2010, is the exciting Solar Dynamics Observatory, or SDO. It returns stunning high-def images at ten wavelengths, including a visual band. Its job is to tie together the puzzles, to monitor the sun's ever-changing magnetism, to stare at flares as they emit their brilliant X-ray flashes, to watch as pulses on the solar surface send sound waves through the interior, and to monitor the sun in the extreme UV that can change its output a thousandfold in minutes, heating up and thickening our own atmosphere in the process.

So while our electrical grid has not been sufficiently hardened to withstand a Carrington event, at least we can see the threat as it's happening. Which is a good thing, since we rely more heavily than ever on electronics, transmission lines, satellites, and jetliners flying polar routes, where solar radiation is strongest. Last year, jetliners made seven thousand such polar flights; thirty-five years ago there were only a dozen. Such vulnerabilities were of course unknown to Richard Carrington, who died at age forty-nine, three decades before the first Wright airplane took off.

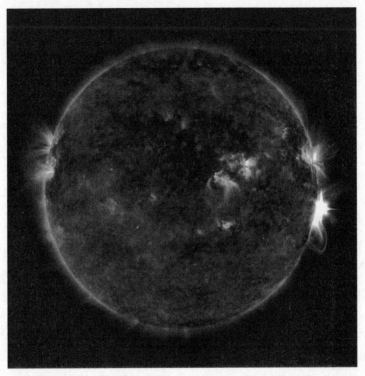

A major solar flare erupts on the right side of the sun on September 10, 2017. Had it occurred near the center, it could have caused problems here on Earth. *(NASA/ SDO/Goddard)*

Maybe the next Carrington event won't happen for another century. Let's hope that's the case. Our brave new world has not included many brave new safeguards against the strongest solar storms — like that prototypical maelstrom Carrington accidentally witnessed when he was thirty-three years old.

CHAPTER 14

CAN WE TRUST SPACE ITSELF?

Not too long ago, the word *vacuum* meant "nothing" to physicists. It was the absence of anything. What could be simpler or more benign?

That's changed for keeps. There's good evidence the universe's vast tracts of emptiness seethe with unimaginable power. And, according to many theorists, that emptiness may someday turn on us and actually destroy the universe. Talk about the ultimate cataclysm.

In a way, it all starts with the ancient Greeks, who hated the notion of a vacuum. Their argument was semantic, stemming from their love of logic. How could there *be* a *vacuum,* they reasoned, when the concepts of *be* and *vacuum* are contradictory? If a vacuum is nothingness, well, there can't *be* nothing. Later, theologians jumped into the fray to support the Greeks. Their stance was that a vacuum would mean that God or His creation of nature was totally absent, and since God by definition was omnipresent, something had to give, and it wasn't going to be God. It was the notion of emptiness.

Those silly Greeks, this author thought back in the 1960s when he first studied physics in college. *Of course you can create a vacuum. Just evacuate all the air out of a bell jar, every last molecule, and voilà.* Laboratory vacuums still had a few molecules, but so what? That hardly mattered.

Turns out the Greeks were right. First, no matter how good the vacuum, its space is still penetrated by some heat from the vacuum's walls or its environment. Since energy and mass are fundamentally the same, those waves zipping through all of space mean that you can't ever have a true vacuum.

But that's small potatoes compared to this next item. Heisenberg's uncertainty principle says that a vacuum shouldn't exist, and other theorists argued that the vacuum of space should be filled with a bizarre sort of quantum energy.

Then came 1979's Big Bang Inflation theory along with a hypothetical new "false vacuum" out of which springs gravity, matter, and, eventually, everything else. Still earlier, quantum electrodynamics (QED) theory had predicted that even total emptiness has properties. Confusing? Maybe it's time to review nothing.

Too bad we have to do this here. Earth is not a great place to contemplate vacuums. Our neighborhood is unusually dense; human fingertips boast a trillion quintillion atoms apiece. And each of these, a single fingertip, contains nuclei made of protons and neutrons that are so dense, one sugar cube of them alone, with all the vacant space removed, would weigh five hundred million tons.

That you can pick up a fork without a forklift suggests that vast emptiness envelops and surrounds all these nuclear particles. If each atom were a vacant football stadium, its nucleus would be a single grain of salt at midfield. And even entire atoms (with the electron orbitals represented by the stadium seats in our model) are smaller than most people imagine. An atom has the same proportional size to an apple that the apple has to planet Earth.

Thus, solid matter is actually quite empty — empty enough that you could easily shove your hand through a table. The palpable pressure preventing this comes not from your atoms meeting the table's but from its electrical fields encountering yours. *We never feel solids. We feel energy fields.*

The tiny particles within all this empty space jiggle wildly. Protons throughout your body squirm around at about sixty thousand miles per second; the much larger molecules they inhabit swarm around at about two hundred and fifty miles per hour from ordinary body heat. And things are not much calmer in the near-perfect frozen vacuums of distant space, places with barely any atomic fragments. QED says that "virtual particles" continuously snap and crackle in and out of this nothingness, though this has never been observed. Moreover, nearby energy fields might let evanescent particles borrow some energy and become permanently real (like Pinocchio). Indeed, the Big Bang, many believe, was a larger version of this phenomenon — a quantum fluctuation.

Unlike ordinary well-behaved vacuums, a theoretical unstable "false vacuum," some cosmologists think, exploded in an expanding frenzy of energy and gravity to create the entire universe. Leftover bits of false vacuum may forever seed countless other universes as well. Supporting this hypothesis, an Anglo-Australian team announced in 1998 that their study of galaxy clustering confirmed that the cosmos started accelerating its expansion when it was half its present age. This is now attributed to runaway vacuum energy (also called negative gravity or dark energy) exerting dominance over matter. Once responsible for the inflation that jump-started the cosmic expansion, negative vacuum pressure might thus be manifesting itself anew; all that was needed was for the universe to grow large enough for ordinary gravity to become too diluted to counteract it.

So, case closed? Has the universe's genesis been adequately explained as a free lunch from a simple vacuum fluctuation? This is where we look a bit more deeply to see if we're uncovering cosmic secrets as opposed to playing logic games.

If some theorists consider the universe's matter as positive and its gravity as negative, and the Big Bang created both in equal quantities, well, in one sense, the positive and negative cancel each other and the universe's sum is zero. Therefore no violation of getting something from nothing existed in the Big Bang, since nothing was ultimately created — and it all zeroes out.

Some theorists have thus redefined various universe components to find a way to say that everything is really nothing, so if the cosmos arose from a vacuum, it doesn't violate anything, because in some sense the universe *is* a vacuum. Of course, you won't be blamed for wondering if this is at all helpful or if anything has actually been explained by this sort of double-talk.

Time will tell if "nothing" can indeed explain everything. But meanwhile, it has been obvious for most of the past century that the emptiness of space is not nothing, and this is where the first faint hint of peril arises, for this change from long-held beliefs can set the stage for the ultimate cataclysm.

The universe does seem to be mostly composed of traditional nothingness. Even here on Earth, the richness around us is an illusion. Remove all the unoccupied space within each atom, and a gathering of the entire human race would take up the volume of yet another mere sugar cube.[1]

For centuries, scientists assumed and trusted that space was filled with a plenum or ether because, just as sound is a "wave" of complex alternating air compressions or similar waves through some other medium, like water or metal, light seemingly required some medium through which to propagate. When the famous Michelson-Morley experiment in 1887 showed that Earth apparently traveled through nothingness, it was a relief to astrophysicists, whose case was nailed by Einstein's first 1905 relativity theory, which showed that light travels happily through a vacuum. The absence of any ether or plenum was deeply appreciated when it came to celestial objects. It hadn't made sense that Earth in its orbit should be whizzing through any substance at 18.5 miles a second without the slightest resistance.

And so emptiness was assumed to rule the universe. Emptiness had been perplexing when it came to figuring out "what light is a wave of," but when scientists learned that light was self-contained and had its own magnetic and electric fields that oscillated without any outside interaction, they could drop the imagined necessity of some supportive medium.

This new elevated status for the void lasted for decades, but gradually scientists came to see that outer space couldn't actually be devoid of all mass, since tiny items hurled away from the sun, like homeless protons and electrons, inhabited such space after all. In the 1950s, the solar physicist Eugene Parker said that the sun sent out a constant stream of disembodied atom fragments, a solar wind, with an average density of three to six atoms per sugar-cube volume. It was substantive enough to push comet tails backward, like airport wind socks, and make them always point away from the sun.

In addition, cosmic rays continually hit our planet, and these are tiny solid particles too. But there's more to space than its particles. For starters, it's permeated by magnetic and electrical fields. And photons, or particles of light. And since light's energy had a mass equivalent, according to Einstein, there was no true vacuum anywhere. The reign of nothingness thus came and went in the twentieth century.

Then there's the small matter of neutrinos. These are fundamental particles that weigh almost nothing and thus travel at very nearly the speed of light.[2] Floods of them are born in nuclear reactions, and the sun's core sends out more than we can imagine. About four trillion of them fly through each of your eyeballs every second.

Oddly enough, their numbers do not diminish at night when the sun is absent. That's because they are electrically neutral, so they're not bothered by electrical or magnetic fields, and they do not normally interact with matter at all. In fact, they easily fly through the entire Earth as if it were no more substantial than fog. By day, the neutrinos enter your head and exit your feet. By night, just as many come through you from below, after having penetrated all eight thousand miles of the Earth in one-twentieth of a second. They then exit your shoulders and hair, heading upward. It would take a wall of lead half a light-year thick to stop a neutrino. The body of Earth totally shields you from cosmic rays, which hit you from above but not from below. It doesn't shield you from neutrinos.

When these ever-present neutrinos are factored in, along with the energy fields and photons from all manner of invisible forms of light arising from all over the universe, it's clear that a lot of energy is present, even if it all weighs little or nothing.

Most intriguing is space's omnipresent *vacuum energy*. Also called zero-point energy, it lurks everywhere and it's substantial. Estimates vary, but each empty mayonnaise jar of space may contain enough power to boil off the Pacific Ocean in one second.

Much experimental evidence shows that virtual particles — things like electrons and antimatter positrons — snap, crackle, and pop out of nothingness everywhere and all the time. Each particle typically exists for just a billionth of a trillionth of a second, then vanishes. If there's an energy field around, a subatomic particle can use some of the energy to remain in existence. Thus, things perpetually materialize out of that quantum vacuum. It's as if the entire cosmos, despite the appearance of emptiness, seethes with so much energy it can barely contain itself. This is probably the so-called dark energy that's making the visible cosmos expand. If this quality of space was what caused the Big Bang, then the universe is still banging, all thanks to its "empty" space.

Most physicists now believe that this underlying quantum foam, along with its associated vacuum energy, has an unimaginable strength. Estimates of the power in each small bit of seemingly empty space vary enormously. Indeed, the difference between the value actually observed and what is predicted is so enormous, it's been called the "vacuum catastrophe." It certainly illustrates how physics is still in its toddler stage as scientists try to understand this underlying power intimately embedded within seeming nothingness.

That's why it's still too early to take seriously the hopes of some dreamers to exploit this vacuum energy to give the world unlimited power. One big problem is that this energy exists everywhere equally (which is why we don't sense it or detect it). As a rule, energy flows from a place of greater energy to a place of lesser energy. So how would you set up a condition that had less energy than that which is everywhere? How could you make it come to you?

The closest we get is to chill matter to absolute zero, −459.67 degrees Fahrenheit, where all molecular motion stops. Then and only then, at absolute zero, is the environment at parity with this all-pervasive power, which is why it's also called zero-point energy. There's evidence that this hidden fount of energy does start to show itself then. How could helium still be a liquid at absolute zero if it weren't receiving a bit of energy that kept it from freezing solid?

In short, zero-point energy does show itself, but only when all other energy is absent.

To get this limitless quantum-foam energy to flow to you, you'd have to somehow create below-absolute-zero conditions. You'd have to make molecules move slower than "stopped." (If you have any ideas, we scientists are listening.)

Harder to grasp is an entirely different aspect of emptiness, one that has changed space from logical to enigmatic. Since the late 1990s, experiments have confirmed the reality of entanglement. Here, two bits of light or actual physical objects, even clumps of material that were created together, fly off and live separate lives but are always "aware" of the other's status. If one is measured or observed, its twin knows this is happening and instantaneously assumes the guise of a particle or bit of light with complementary properties. This "information" propagates through empty space with no time lag, even if the twins are on opposite sides of the galaxy. In short, space is penetrated instantaneously, in zero time, no matter the distance.

This strongly suggests that the gap between bodies is not real on some level. Emptiness is not what we once assumed it to be. If far-apart objects can be in instantaneous contact no matter the distance, what does this say about any practical meaning of space in the sense of separation?

And wait—the problem of visualizing space gets even worse. Einstein's theory of relativity shows that space is not a constant and therefore is not inherently tangible. High-speed travel makes intervening space dramatically shrink. Thus, when you step out under the stars, you may marvel at their distances and the universe's vast spaces. But it has been shown repeatedly that this seeming separation between yourself and anything else is subject to point of view—what Einstein called a reference frame—and therefore has no *inherent* bedrock reality. This doesn't by itself negate space; it merely makes it tentative or a component of space-time that in itself has no substantive existence. It's like removing the tick or the tock from a stopwatch and then wondering why you no longer possess a device that keeps time. Space-time is merely a mathematical framework that reveals how objects or bits of light behave or interact.

If we lived on a world with a very strong gravitational field or if we traveled outbound at a high speed, the night's stars would lie at entirely different distances. If we headed toward the winter star Procyon at 99 percent of light speed, we would find that it was actually less than two light-years away, not the eleven light-years we'd previously measured it to be. If we crossed a living room twenty-one feet in length going at that speed, every instrument and perception would show that it was actually now three feet long. And if we could move at 99.9999999 percent of light speed, which is perfectly allowable by the laws of physics, the living room would now be barely larger than the period at the end of this sentence. Space would have changed to nearly nothing. Where, then, is the supposedly trustworthy space matrix, the grid within which we observe the stars and galaxies?

So Einstein showed that distances mutate according to the conditions surrounding the observer. Bottom line is that the universe doesn't have a fixed size. It is neither small, nor large, nor enormous. In truth, the cosmos is fundamentally sizeless.

Quantum theory entered the picture to cast serious doubt on whether even distant individual items are truly separated. So the issue of seeming nothingness is not merely the gap between objects; it seethes with power. And, as if this weren't enough, it can almost magically alter its size.

There's also a mystical aspect of space. Since several of these variations depend on the observer, it's obvious that awareness or consciousness must somehow be connected with all this. Indeed, the unity or correlativeness of nature and the observer is a central theme of the biocentrism model, discussed extensively in books that bear that term, co-written by your present author and Robert Lanza, MD. In it, the authors make clear that while most people regard space as sort of a vast container with no walls, it's actually part of the mind/awareness/consciousness matrix that permeates the cosmos with various measurable effects.

If all of this makes your head spin — as it does this author's — consider the bedrock fact that the cosmos is inarguably inflating. Massive objects such as galaxies cannot be what's accelerating, simply because nothing of mass can reach light speed, and in the actual cosmos, most of existence is zooming beyond that speed. Thus, following the advice of Sherlock Holmes and having eliminated the impossible, we're left to conclude that the space itself is inflating.

A century ago, when space was mere nothingness, people would have been flummoxed, because how can nothingness do anything? What sense does it make to declare that, on its own, space is growing larger? But now that we know that what we call space is a matrix of energy fields, photons, neutrinos, vacuum energy, vacuum foam, and perhaps consciousness as well, this amalgam is not nothing. It is something, even if our senses cannot perceive it.

The invisibility aspect shouldn't be any kind of stopper. We can't see Earth's gravity or its magnetosphere or microwaves, and they're real too. Anyway, the wild motion of every single galaxy cluster in the universe is proof that something powerful is happening to blow it all apart. That's our present reality, and it's inarguable.

The issue is—what will space do next?

It's important to remember that even a vacuum devoid of a single particle still contains energy. We know this thanks to Einstein's $E=mc^2$, which specifies the equivalence of mass and energy. Beyond the omnipresent electromagnetic fields, with photons of infrared and other forms of invisible light zooming through the cosmos from all the galaxies and their stellar contents, there are also quantum fluctuations caused by the Heisenberg uncertainty principle. This makes virtual particles pop briefly in and out of existence, and all of this constitutes power or energy, that zero-point energy whose effects are best detected when atoms come to a stop at absolute zero.

Normally, temperature or heat simply means the motion of atoms, so atoms chilled to -459.67 are at absolute zero because they've stopped jiggling or moving. But even here, the quantum fluctuations continue, and evanescent particles still arise and subside, and this constitutes its own energy. We can call this the underlying vacuum energy or, alternatively, Z-point energy. This plays a big role in clarifying why empty space is not true nothingness—and all this, in case you lost track, is still the setup for our denouement, the possible enormous hazard awaiting us someday.

It's fair to ask how scientists can be sure of this. Let's review the answer provided by the Dutch physicist Hendrik Casimir in 1948. He suggested hanging two metal plates in a vacuum chamber so that there was only a slight space between them—about twenty times the diameter of an atom. If space is just emptiness, they should hang freely. But if space has energy—in this case, the quantum energy exhibited by potential particles—this should push the plates from the outside, and yet the limited space inside would be insufficient to allow any significant pushback. Casimir predicted that the plates would be pressed together by about the force exerted by one atmospheric pressure.

When such experiments were performed, the plates *did* slam shut together. The Casimir effect was real. And with it came the proof that space is not mere emptiness. Indeed, its power is almost unimaginable.

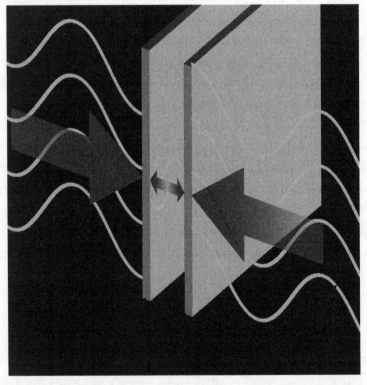

The Casimir effect, theorized by Hendrik Casimir in 1948, violently pushes two hanging metal plates together. It does so because the narrow space between the plates cannot resist the quantum fluctuation forces that pervade the universe's empty space. *(Wikimedia Commons)*

Currently, we observe space to be expanding with, apparently, constantly increasing force. But the putative cause, termed *dark energy*, is still mysterious. Some theorists believe that it can possibly reverse itself over time, which would lead to the cosmos shrinking, at first slowly, and then with ever greater speed. Such an event is entirely speculative and wouldn't be worth our time to consider if it wasn't in some way possible. But if it does happen, it would be goodbye to everything.

The concept of such an eventual Big Crunch got a boost in 2014 with a paper in *Physical Review Letters* by two theorists who argued that the vacuum energy currently driving the universe's accelerated expansion should logically reach a maximum effect just before it turns around, which should occur about ten billion years after the Big Bang—or, very roughly, now. They offered a "simple mechanism" to bring about this collapse, due to a "scalar field whose potential is linear and becomes negative, providing the [required] negative energy density." These physicists believe that the collapse of the universe is "imminent."

Once it began, we'd still have billions of years before we started ignoring prudent dietary restrictions and otherwise got ready for oblivion. One visible change indicating that this slow-motion cataclysm was indeed under way would involve the large-scale structures all around us in space. Currently, images of early galaxies seen through large telescopes reveal that some ten to twelve billion years ago, the cosmological structure was highly uniform; when you examine a sample in any direction, it looks very much like the configuration observed in other directions. But such a uniformity would become rarer in a contracting universe. The cosmos would grow increasingly clumpy, with galaxy clusters merging and individual galaxies merging into larger ones. Spiral galaxies would combine with one another until such pinwheels became very unusual. Football-shaped and spheroidal galaxies would be the new normal.

As the cosmos got even smaller, more matter in each of these groupings would collapse into black holes, and then these too would coalesce. The end point would be a single unified black hole, usually termed a Big Crunch singularity.

This would extinguish all life. And yet it wouldn't necessarily be the final chapter. No one could say whether the cosmos might then somehow rebound, with dark energy reasserting its former repulsive property, thus initiating a Big Bounce and a new Big Bang with, perhaps, novel laws of physics so that the ensuing universe would be fresh and original. This would happen if the old, esthetically attractive oscillating-universe paradigm proved to be correct, though none of its original twentieth-century proponents would be around to say "told you so." This scenario jibes beautifully with our cataclysm-followed-by-resurrection motif.

This well-known Big Crunch scenario is not the only way that space itself could become the agent of cataclysm. An antithetical calamity, called the Big Rip, has also gained some support. This idea is taken seriously because of a 2003 paper by Robert R. Caldwell, a theoretical physicist and professor of physics and astronomy at Dartmouth College. His paper suggests that as the universe expands, space-time will begin to tear itself apart. This fly-apart disintegration will happen from the inside out, so that even atomic particles will fall into pieces, along with everything built of them — which includes planets, stars, the sun, and galaxies.

In short, it's the opposite of the Big Crunch, where gravity and dark matter win the battle either because they've been a bit stronger than dark energy all along or because dark energy inherently decreases and thus loses its dominance or even reverses itself.

But that's not what we see today. Instead, dark energy very much appears to be gaining strength as the cosmos grows emptier and larger. As it does so, objects at the boundary of the visible universe, which means the place where galaxies are flying away at the speed of light, zoom ever faster, *which pulls that boundary closer to us.*

Say what? Time out—since any contemplation of these farthest visible places deserves a closer look.

As we saw in chapter 2, the Hubble constant, or rate of the universe's expansion, is fourteen miles per second for each million light-years farther that we look. This means that at 13.8 billion light-years, everything is rushing away at 186,282.4 miles a second, the speed of light. Light from all objects farther away than this cannot ever reach us.

But if all speeds are increasing because of dark energy, then the Hubble constant is not a constant over time. This inflation rate will someday grow to fifteen miles per second per million light-years of distance. Then sixteen, then seventeen, and eventually to a hundred miles a second and more. All the while, this would make the place at which the recession speed is 186,282 miles per second move gradually closer to us. Put another way, while the overall universe kept getting bigger, the *visible universe* would be continually shrinking.

As time went on, more and more of the cosmos would lie over the horizon and become invisible. As the visible universe shrank, the fly-apart tendency would come closer, like wolves circling a home. Currently, galaxies are held firmly in their shapes by gravity, and their component stars do not grow more separated with time. Even galaxy clusters keep their members at the same stable distance. Right now, what's happening is merely that each galaxy *cluster* is becoming more separated from other galaxy clusters.

But if dark energy grows in strength, the clusters themselves will start to be pulled apart. Then, predicts the Big Rip scenario, galaxies will grow more swollen until they disperse their contents like piñatas. What these theorists are saying is that the process keeps going until even the local space-time, including the pieces of it in our bodies and brains, rip apart, the way excessive partying already feels.

As space-time itself tears apart, everything we know disintegrates.

We might wish to vote for which cataclysm we prefer, a space-time rip or a space-time crunch. Either would constitute the ultimate cataclysm. Neither would offer any place of refuge. Happily, the Big Rip theorists predict that it won't happen for another twenty-two billion years. Since the sun itself will collapse into a useless white dwarf in less than a quarter of that time, it probably shouldn't keep us up at night.

CHAPTER 15

THE FINAL SUPERNOVA

Our own Milky Way has not had a supernova in over four hundred years. Not a single brilliant new starlike point has lit up the heavens and cast shadows since just before the telescope was invented. Those wild halcyon supernova years in the eleventh, sixteenth, and early seventeenth centuries are for the history books alone.

Or so we thought. One strange night in February of 1987, our dwarf companion galaxy, the Large Magellanic Cloud, or LMC, had a star go ka-plooey with enough brilliance to wreck its poor solar system and send a flood of detritus through human bodies 166,000 light-years away. And enough brilliance to be seen by the naked eye.

Are you kidding? A naked-eye supernova after four centuries?

The author offered the chance for a peek to a bunch of friends, and we booked a flight to Costa Rica, which was far enough south for the exploding star to appear well above the horizon. Sure enough, there it was, neither challengingly faint nor eye-catchingly bright. It had been slowly brightening for two months and now matched the luminosity of the Little Dipper stars, meaning it was half the second-magnitude brightness of Orion's belt. This supernova was magnitude 3. In keeping with long-standing tradition, it was called 1987A to signify the first observed supernova in that year, and until it faded a year later, it remained the sole naked-eye supernova since the lifetimes of Kepler and Galileo. It's still the only supernova that anyone alive today can claim to have seen without optical equipment.

More excitement was in store. Each year your author leads an astronomy group to Chile's marvelous Atacama Desert, and one afternoon each year we spend a few hours at the remote cutting-edge observatory run by the Carnegie Institution on a mountaintop called Las Campanas. The past and now present directors have been gracious in hosting an astronomy tour led by this astronomy journalist, even though the facility has no visitors' center and no established procedure for putting up with nonworking astrophysicists. This is how we met Oscar Duhalde more than twenty years ago. He's there nearly every night and runs the sophisticated equipment for astrophysicists who have been awarded time on one of their giant instruments, which include the famous twin 6.5-meter behemoths named the Magellan telescopes. And one night a few years ago, one of those astronomers casually said, "Did you know that Oscar was the one who discovered the 1987 supernova?"

All the guests in that forty-person group simultaneously swiveled their heads from him to Oscar, as if they were at a tennis match. Oscar? The nicest, most low-key, most unassuming man in the world? Our Oscar? He, of all the seven billion humans on the planet, was the first person in four hundred years to see a naked-eye supernova? And it had happened right here? Then why wasn't his name in all the books?

He told us his story. On February 24, 1987, while doing his technical and instrument tasks, he stepped outside to walk to another observatory building. He looked up into the amazingly clear Chilean skies, where the Milky Way is typically bright enough to cast shadows. He glanced casually at the Large Magellanic Cloud, which was high up at that time of year in Chile, though it never clears the horizon in the mainland United States, and he instantly saw something so weird it was next to impossible.

There is no bright star or even dim star in front of that little nebulous patch that is the very nearest galaxy to ours. But that night there was. Oscar knew the stars intimately, the way your author and perhaps only 0.001 percent of the world does, and so he instantly recognized that this new star had to be a supernova. The entire process of spotting and identifying the first naked-eye supernova in four centuries had required no more than ten seconds.

Oscar returned to the big dome and told people. One of them, the Canadian astronomer Ian Shelton, was just finishing taking an image, by amazing coincidence, of the LMC, and he looked at the plate. Yes, a new star was there, and he was the first to capture it. So Shelton sent someone to make the two-hour drive down the eight-thousand-foot mountain to La Serena, where he could place a phone call and report it. Since Shelton was the first to report it, he is usually credited with being the discoverer of 1987A.

But everyone at the Carnegie Las Campanas Observatory knew who had seen it first. So, though it took a few years, you'll now find most literature lists both men as co-discoverers.

But the fireworks were just beginning.

Using various orbiting detectors, the Hubble Space Telescope, and sophisticated new terrestrial telescopes like the newly built ALMA array in the Atacama Desert just a few hundred miles north of the discovery mountain, astronomers feasted on a wealth of data from this nearest supernova since the early seventeenth century. They saw X-rays streaming off to collide with surrounding material that had been previously shed by the unfortunate ex-star, making it glow like holiday decorations. They detected a mass of neutrinos that actually arrived ahead of the visible supernova. They scoured pre-nova photographs and identified the star that had exploded — the first time an exploding star could be pinpointed and examined in an image taken before the cataclysmic detonation. Turned out, it was an enormous blue supergiant weighing eighteen times more than our sun with the catchy name of Sanduleak −69°202.

Supernova 1987A as it looked a half a year later when a lethal shock wave of ultra-violet light smashed into a ring of material in its vicinity. These gassy knots had been shed into space twenty thousand years earlier by the now-destroyed star during a prior outburst of lesser violence. The type 2 supernova was the first naked-eye supernova in more than four hundred years. *(Hubble Space Telescope / NASA)*

Those neutrinos were a really cool thing. Theory had insisted that this type of supernova should produce a flood of those strange fundamental weightless (or perhaps *near*-massless) particles, and by 1987 the world had two separate neutrino-detecting experiments in operation. The fact that the neutrinos were detected before the visible light was observed supplied key information that helped settle the old issue of whether neutrinos had mass. If they possessed even a little heft, they might solve the longtime riddle about the universe's missing mass and whether our cosmos will someday stop blowing itself up and instead start collapsing simply because there are so many neutrinos and each is contributing to the collective gravitational pull. But with any mass at all, they couldn't travel at light speed. So now, finally, supernova 1987A released neutrinos and light at the same time (although theorists said that the neutrinos actually had a few hours' head start). Bottom line: If they had mass, they'd arrive at Earth hours, days, months, years, or even centuries after the visible light of the explosion, depending on their individual weight. But it so happened that they beat the light here by a few hours, after both had zoomed through space for 166,000 years.

So, puzzle solved: Neutrinos have no significant mass. This was the first belated Valentine's Day present SN 1987A delivered to astrophysicists, and the good-news section of the good news/bad news story of that solar-system-destroying cataclysm.

Because neutrinos normally don't interact with the baryonic matter that makes up our planet's elements and their constituent subatomic particles, a signal that a neutrino has arrived involves indirect detections. Even then, trillions of neutrinos must arrive before a single one can change one element to a different one, which is how all that elaborate underground neutrino-finding equipment operates.

But even before the extreme light of a hundred million suns had started to leave the fragmenting Sanduleak star, and thus hours before Oscar saw that new little star in the LMC, neutrinos from SN 1987A were penetrating the Earth. Four trillion neutrinos per second passed unfelt through each eyeball of the planet's nearly six and a half billion humans. But as they penetrated the ground, a few were finally detected.

Lying deep in the Kamioka zinc mine in Japan and in the Morton salt mine under Lake Erie are two massive pools of black water designed to detect flashes of light from a rare neutrino effect. At 7:35:35 universal time on February 23, two hours before Oscar noticed the optical light arriving on earth, the neutrinos reached Japan's Kamioka Observatory's Institute for Cosmic Ray Research, which can detect underground neutrinos by looking at Cherenkov radiation, a blue glow caused by anything traveling faster than the speed of light.[1]

The Kamiokande II detector observed eleven neutrinos from the supernova. (Some sources say that Kamiokande detected not eleven but twelve neutrinos from SN 1987A. The author is tempted to split the difference and say around eleven and a half neutrinos.)

These near-dozen neutrinos detected in that Japanese underground water pool and the eight more found by detectors in that old Morton salt mine underneath Lake Erie were historic. The chance of a neutrino interacting with an atom is so incredibly minuscule that detecting those nineteen or twenty meant that at that moment, ten billion neutrinos from the supernova must have been simultaneously zooming through every fingernail-size area of every earthly human and animal, marking the instant 166,000 years ago when the star's core bounced back from its collapsing layers. This happened just before a blinding brilliance left the star to begin a 1,600-century journey to Earth.

Even if all this unfolded outside the awareness of nearly the entire human race, these astounding detection capabilities deserve a salute. These new methods in which science successfully pinned down the mayhem happening in our companion galaxy were an astounding change from Kepler's Star of 1604—the previous time humans tried to make sense of the naked-eye death of a massive sun.

CHAPTER 16

THE 2017 KILONOVA

A distant celestial explosion is rarely the lead story in the major media, but it was in October of 2017. That's when TV newscasters and print journalists struggled to convey an astounding but technically challenging event to a public that, according to surveys, scarcely recalled the meaning of *electron*.

The explosion resulted from the collision between the densest visible objects in the universe. It had happened on August 17, 2017, in NGC 4993, a galaxy invisible to the naked eye as well as permanently below the horizon for people in the United States and Europe.

The violence was not nearly as bright as a supernova. It was not nearby. Why, then, was it such a big deal?

Gold. Smash together two planet Earths' worth of superheavy atoms, and some will stick together to form matter so dense, it's rarely seen anywhere in the galaxy. The Big Bang famously created the lightweight, cheap, common elements. Hydrogen, helium, lithium, and their isotopes (versions with extra neutrons in their nuclei) were the only elements to be seen after the smoke from the Big Bang had cleared.

By early in the twentieth century, physicists realized that star cores were factories that slowly changed these elements into denser ones like carbon, nitrogen, and even iron. But logic reached its limits with iron.

Converting lighter elements to heavier ones—nuclear fusion—produces energy, and lots of it. In the sun, two hydrogen atoms collide and fuse together ninety-two billion billion billion billion times a second. As a star ages and uses up its hydrogen fuel, it turns to burning helium, then to burning increasingly heavy elements. But the buck stops when it comes to iron. Iron atoms fusing together do not produce energy. On the contrary, the process requires energy. So if the idea is for a star to shine, iron marks the end of the line.

In that case, why does the cosmos possess still heavier atoms like iodine, lead, and platinum? It was a puzzle until Fred Hoyle came up with a solution in 1954, which he published in a groundbreaking *Astrophysical Journal* article, "On Nuclear Reactions Occurring in Very Hot Stars. I. The Synthesis of Elements from Carbon to Nickel."

What if it didn't matter that iron-fusing couldn't produce energy? What if energy was not needed? What if the energy was already present and otherwise would just go to waste? A type 2 supernova's fantastic temperature would fuse still heavier atoms when the collapsing, imploding layers of such an old massive star hit the incompressible core, producing an astonishing shock wave that rebounded outward.

One extremely strange kind of star would start off pre-crammed with heavier elements, the fuel for forging the heaviest elements of all, things like gold and uranium. All that was needed was a final event—for example, a high-speed contact between two such stars.

When the news media learned of it that autumn day in 2017, they had their story. This collision had probably created as much gold as the weight of fifty planet Earths, and then the kinetic energy hurled it outward, where it might someday form gleaming veins of ore on some newly hatched alien planet. What an image!

The colliding objects were neutron stars, which are the ultra-compressed central cores that are often the sole remnant of supernovas. Neutron stars don't get much mention these days despite being fairly new to us, having been discovered only in the 1960s. To the public, they don't seem as morbidly fascinating as black holes. They're not the densest time-and-space-warping entities in the cosmos, just the second densest. The media shy away from runners-up and wannabes, while fashion and the bandwagon effect may also contribute to their obscurity. A shame. It's like ignoring a lizard man sitting next to you at a concert just because a dinosaur is running through the parking lot. Sure, a neutron star may be a runner-up in the Weirdly Dense Objects competition, but at least it's visible. It shines, it's *there,* it has a visual presence, unlike black holes.

Again, a neutron star's birth is a unique blend of the violent and the esoteric. It's the end product of that rarest kind of star, the supermassive blue sun; its core grows too wimpy to support all the gravitational weight of the outer layers, so the star collapses, heats up in the process, and then ignites in the kind of type 2 supernova we first explored in chapter 5. Much of the star is blown to bits, its fragments hurled outward into space. But some of the material keeps imploding until all that's left is the already heavy core, now even heavier.

In a more typical star that weighs the same as our sun or less, the old-age collapse stops because of electron-degeneracy pressure, and the collapsed ball comes to a sudden stop when it's the size of Earth. This is weird enough, since every teaspoon of this material outweighs a loaded cement truck. It gets no denser because of the quantum fact that you can't have electrons occupying the same spot. They need some elbow room.

But if the star is so massive that its gravity is crazy-intense, this degeneracy pressure won't do the job. Electrons are pressed into protons to form neutrons, and the star's collapse doesn't stop until the whole thing is a dense ball of neutrons, making *neutron star* a literal and descriptive term, not just some sort of whimsical metaphor. It's really essentially a single neutron, but instead of being subatomic, it's twelve miles wide. And like every neutron in your body, it retains that particle's unbelievable density.

Collapse usually causes a star's spin to increase, and many neutron stars spin several times a second. Like a lighthouse, the star can deliver quick bursts of energy with every turn in a wide spectrum of wavelengths. If Earth happens to lie in the path of these pulses, then the neutron star is called a pulsar.

We usually think of star surfaces as gassy and unsupportive. The sun's vaporous visible photosphere is much less dense than water. But a neutron star's surface is a hundred thousand trillion times denser than steel.

Sometimes, the neutron star is a member of a binary-star system. After all, more than half the stars in the universe are not single, like our sun, but members of a two-, three-, even four-sun system, all revolving around their common center of gravity. This was the situation in a spiral arm on the outskirts of the galaxy NGC 4993, lying far beyond the stars that form the southern constellation of Hydra the Serpent. Astronomers have observed neutron stars orbiting black holes or normal stars, but in this case two neutron stars orbited each other, circling their gravity's barycenter, the seesaw balancing point between them.

For years, astronomers had observed strange transient streams of high-energy gamma rays radiating from one or another seemingly blank spot of sky, and theorists had suspected neutron-star collisions for years but could never catch them in the act. All of this changed in 2017, and what made it really special was that the right instrumentation was poised and ready.

First, the energy was detected by the new LIGO instrument, an optical assembly with bouncing laser beams designed to register the slightest changing distance between its mirrors and thus signal if space-time itself was warping. Then, two seconds later, the orbiting Fermi Gamma-Ray Space Telescope detected a two-second burst of gamma rays, an indication of extreme cosmic violence.

Telescopes in Chile, where the galaxy was above the horizon, and, soon, the Hubble Space Telescope photographed a new speck of light on the outskirts of the distant galaxy.

The tiny neutron stars, orbiting closer and closer to each other for years, had increased in speed until they were now spinning around their center of gravity a thousand times a second, with speeds that were a healthy fraction of the speed of light.

Unimaginably intense tidal forces did the impossible and squeezed each ultra-solid seemingly unbendable star into a football shape, rupturing each at its rotation poles and spewing ultradense star-stuff into space at high speed. As the twin pulsars collided with the few surrounding atoms and with themselves, they formed radio-active atoms that created the 8,000-degree heat of the visible fire-ball seen suddenly through earthly telescopes while emitting X-rays and gamma rays on their own. It would have been lethal to any and all planets and their inhabitants in that solar system.

So the cataclysm was noteworthy not because of any unprece-dented brightness or even because new elements were born before our eyes—supernovas do the same thing. Rather, it was a mile-stone moment in science history. It was the very first collision between neutron stars ever directly witnessed. It was also the first time that any visible object was observed as it warped the fabric of space-time, since the few previous LIGO events had been caused by invisible black-hole collisions.

It showed us, once again, how new materials arise from the extreme cataclysmic violence caused by some of the oldest and strangest objects of the universe. And for public relations purposes, it didn't hurt that this time, the physical result was the creation of more gold than the weight of our entire planet.

PART II

CATACLYSMS OF EARTH

CHAPTER 17

THE OXYGEN HOLOCAUST

Five times in Earth's relatively recent history—meaning the period when multicellular animals occupied the oceans and/or the surface—at least half of all life-forms were suddenly obliterated. The most recent and famous of these annihilating events was caused by our planet's collision with a giant comet or asteroid, as we'll see in chapter 19.[1]

Not that every cataclysm is a bad thing. Perhaps none are, in the fullness of time. When violent upheavals erase the global stage, like dust in a car wash, something else follows in the niches that appear. To date, those new conditions, populated by new forms of life, have been more than merely interesting; they've brought us to our present situation. It is for this reason that major disasters that have befallen a largely uninhabited or single-cell-populated Earth will similarly be ignored in this book—especially since the record of such very early violence is now buried so deep it is hard to determine what happened with any precision.

Instead we begin with the earliest known cataclysm after the moon's creation, one that very much affected the future of the planet and that was most central for the appearance of us multi-celled creatures. We're talking about the GOE.

It's the great oxygenation event. It's also commonly called the oxygen holocaust, the oxygen crisis, the oxygen revolution, and the oxygen catastrophe.

These may seem like unduly negative labels since the starring element isn't one of the notorious bad guys like arsenic. You've likely not been surveyed about your favorite element by one of those robocall polls conducted just as you're sitting down to dinner, and it's hard to say whether most folks would choose oxygen over, say, gold or platinum. But it's the only one of the ninety-two natural elements you can purchase at a coin-operated or card-swipe booth in places like Tokyo.

Our planet, like all others in the known universe, had little or no oxygen to begin with, even though it's the third-most-common element in the cosmos. But free oxygen — the gaseous element sitting alone or else combining with itself as O_2 — is rare on planets for a good reason. Its electron configuration makes it unusually fond of sharing electrons. At the slightest provocation, it will link with another element in a process called oxidation. Its favorite target is the universe's most abundant element, hydrogen. When it connects up to form H_2O, it creates the most common compound in the cosmos (a compound, you'll recall from high school, is a substance made of two or more elements).

So when life began on Earth, it had to make do with no free oxygen. In that original matrix 3.9 billion years ago, the creatures that arose — such as cyanobacteria that lived abundantly in the oceans and probably on land too — broke water into its constituent hydrogens and oxygen and thus released free oxygen.

At first only prokaryotic organisms created oxygen as a waste product. Later, eukaryotic creatures did too.[2] But the rare, precious liberated oxygen wasn't given much of a chance to be a free-floating part of the atmosphere. Instead it was quickly snatched up by such greedily reducing elements as iron. Millions and millions of years passed while Earth's silicon, iron, hydrogen, and carbon slowly turned into silicon dioxide (quartz and sand), iron oxide (rust), hydrogen hydroxide (water), and carbon dioxide, which grew more abundant in the atmosphere.

Earth's atmosphere was sent spiraling off balance, even though these were the first steps toward our planet becoming what it is today. You'd think the extra CO_2 would have warmed the planet, since today it's the primary villain in climate change. But the opposite happened. The newly minted oxygen attacked the methane that was initially abundant. Methane is CH_4, and oxygen smashed this larger molecule into fragments as it latched onto its C and its H to form carbon dioxide and water vapor. Since the new CO_2 and H_2O were both less effective greenhouse gases than the CH_4, the methane it replaced, Earth got *cooler*.

Thus, one of the early consequences of the oxygen holocaust was that ice ages began. But even before these took a toll on Earth's life-forms, the oxygen was deadly to the prevailing anaerobic organisms. The cyanobacteria, by unleashing a very un-Darwinian mass homicide/suicide, created the first global cataclysm since the moon's birth, wiping out most of the life on our planet.

Needless to say, this oxygen catastrophe had a plus side; it led to an event that was probably the most dramatic and important in the history of our world. Yet, in an indictment of our education system, if people are asked to name the single most influential occurrence in the planet's history, few are knowledgeable enough to cite the Cambrian explosion. Most people probably don't even know what it is.

The Cambrian explosion might sound like some sort of huge detonation, one more item in our recitation of cataclysms, but it's almost the exact opposite—it's the odd term for the sudden remarkable appearance and proliferation of multicellular animals and plants. Prior to this time, which was about five hundred and fifty million years ago, life was invisible. It was all microscopic. Now a strange, wild profusion of large creatures materialized quickly and globally.

And although the first creatures, like trilobites, looked alien, and the plants were largely different from today's flora (there was no such thing as grass, for example), the planet was now a visibly alive biosphere. The oxygen holocaust had effectively erased a global situation that had endured for hundreds of millions of centuries.

We oxygen-adorers, or at least those of us with a them-versus-us mind-set, might hesitate to label that wipeout of non-oxygen-loving creatures a cataclysm. But you'd feel differently, of course, if you were the final anaerobic organism watching your extended family die off, knowing that this was really the end of an era that had endured for not millions but billions of years.

Enough violin music. Now that the scene has shifted to a planet that is more familiar, we can focus on the cataclysms that were the greatest of all in terms of affecting our kith and kin, the beloved multicellulars.

CHAPTER 18

THE GREATEST MASS EXTINCTION

Chicago keeps returning to our narrative. It was the Chicagoan Edwin Hubble who revealed the expanding universe that made the initial case for our universe's birth violence, the Big Bang. Then two more brains from that city, paleontologists at the University of Chicago, changed the world's thinking about mass extinctions.

J. John Sepkoski Jr. and David M. Raup were U of C colleagues who came to regard the great dinosaur extinction sixty-six million years ago as part of a pattern. Just a couple of years prior to their paradigm-changing 1982 paper, the dinosaur-ending event (now called the C-Pg extinction, which we'll explore in a moment) was seen as a onetime affair, explained as a freak asteroid or comet impact. But Sepkoski and Raup suggested that it was part of a *series* of mass extinctions that formed a deadly blueprint of cataclysms that recurred every twenty-six million years.

Though Sepkoski died in 1999 at age fifty, both researchers lived to see their conclusions proven both right and wrong. Right, because we now know that there were indeed at least five major extinction events. And wrong, because no cataclysm periodicity has been found. The mayhem happens randomly.

In chronological order, these biological disasters were, first, the two Ordovician-Silurian extinctions, usually considered together, which brought 60 to 70 percent of all earthly species to extinction

450 million years ago. Then came the late Devonian extinction, which terminated more than 70 percent of all species over an extended period that may have lasted twenty million years and that finally stopped 360 million years ago.

Then Earth got a break for about a hundred million years before the worst of all extinctions fell on the planet like a ton of bricks. This "Great Dying" 252 million years ago was so nearly curtains for earthly life that we will spend most of this chapter on that single cataclysm.

Some fifty million years after that, the Triassic-Jurassic extinction event drove 70 to 75 percent of all species to extinction. And then our planet enjoyed relative calm for a lengthy 140 million years until the giant asteroid hit us and delivered the most well-known extinction, the dinosaur-ending cataclysm we will explore in the next chapter. (There is likely a sixth extinction at the current time due to the havoc we humans are wreaking on the planet.)

These five great extinctions millions of years ago each devastated the Earth's large, multicelled creatures and global plant life, but we will represent and memorialize them by focusing on two. In the next chapter, we will investigate the K-T event (now called the C-Pg event) because it was the most recent and also because it brought about our species' own dominance, and thus we spokesmammals have a personal retroactive stake in that game and are far from uninterested in its outcome. But first we must explore the other event, selected because it was the worst of the lot and the most dramatic.

The Great Dying 252 million years ago has a bunch of official designations that include the Permian-Triassic, or P-T, event, the great Permian extinction, and the end-Permian extinction.[1]

When the Great Dying was taking place, earthly conditions were anything but fabulous. The terrestrial environment was both hostile and bizarre. For example, the ocean's surface temperature may have reached 104 degrees Fahrenheit, matching that of a modern hot tub. Obviously, not many organisms were going to survive in that, and, indeed, this extinction killed an amazing 96 percent of all marine species. As if that wasn't enough, the carbon dioxide in Earth's air, which is presently a worrisome 404 parts per million, compared to its pre-industrial level of 280 parts per million, reached 2,000 parts per million during the extinction years. It was enough to poison the oceans through hypercapnia, or excess carbon dioxide. There was even enough CO_2 to occasionally bubble out, literally turning the sea into hot soda water.

On top of this, or perhaps because of it, the oceans suffered widespread and probably universal anoxia, or lack of oxygen. Marine organisms were devastated, and ocean creatures that had thrived for hundreds of millions of years suddenly went extinct. The ocean's superhigh CO_2 concentration also acidified the seas globally and devastated coral, sponges, and all creatures that incorporate significant carbonates into their skeletal systems. One reason for the extreme marine-organism susceptibility is that carbon dioxide is twenty-eight times more soluble in water than oxygen is, which is one reason why soda pop manufacturers use CO_2 for fizziness rather than oxygen or some other gas.

But if conditions were lethal in the oceans, they weren't much better on land. Earth experienced a global air temperature increase of fourteen degrees, and 70 percent of all vertebrate species were extinguished. Indeed, the end-Permian event was the only one that produced a mass extinction of insects.

Even the plants suffered despite the greatly enhanced CO_2 level. Nearly all of Earth's trees were killed, as evidenced by the disappearance of tree pollen at the time and the sudden profusion of spores, indicating that the rotting biomass provided a feast for fungi.

And, unlike the later asteroid-caused extinction, following which the biosphere repaired itself, albeit with a new mix of plants and animals, after this event earthly life was very slow to recover. Some evidence points to a ten-million-year recovery time, which is close to unimaginably lethargic, but it certainly required at least four million years. The reason for this tenacious lethality probably began with the increased CO_2 and anoxia of the ocean, which could not readily heal itself from such conditions, and that anti-life ambience was augmented by the extreme and persistent surface heat. It was definitely a close call for our biosphere, the cataclysm of cataclysms.

Researchers desperately want to know what caused this greatest of all mass extinctions. But the evidence here, unlike in the more recent C-Pg event, is scant, contradictory, and puzzling. And there are many plausible mechanisms that could have done the job singly or, with particular virulence, in combination with one another.

For example, rapid mutations of plant spores from that time indicate that a strangely high ultraviolet flux was pounding the Earth's surface. This could have come from extreme solar activity, but that wouldn't explain the bizarre carbon dioxide levels.

There are four likely causes of the Permian extinction. Since none has been unequivocally proven, you, dear reader, are free to pick your poison.

The first is a good old-fashioned extraterrestrial impact, similar to the C-Pg dinosaur-ending event. No decisively dated crater or global iridium deposit layer is available to prove that this is the correct explanation. But there are vast areas of shocked quartz in Australia and Antarctica whose origins are not inconsistent with an impact 252 million years ago.

Since more than two-thirds of our planet is covered by oceans and may have been back then as well, the odds are two to one that the impact would have been underwater. Unfortunately, no crater from that time would likely still be detectable, since the ocean floor completely remakes itself every two hundred million years from seabed-spreading and tectonic-plate activity. The Permian extinction was simply too long ago to have left a visible scar.

More intriguing are the Siberian Traps, the odd name for a

series of ultrapowerful, interlinked volcanoes that definitely erupted around the time of the extinction. These would have thrown huge amounts of ash and carbon dioxide into the air and could have been the source of the extreme atmospheric and eventual marine carbonization of that period. The strongest anti-life evidence of that era is of widespread marine anoxia and hypercapnia, as well as oceanic euxinia, which means poisoning by hydrogen sulfide, the odious stink-bomb compound. All of these could logically follow from the excessive gas emissions that would be expected from the Siberian Traps.

Another odd feature from 250 million years ago is an odd iso-tope ratio of carbon-13 to carbon-12, so an event that could have produced that ratio would be particularly compelling—and the super-eruptions of the Siberian Traps would qualify.

An odder alternative explanation comes from strange living organisms. Research in 2014 suggested the P-T extinction arose because of a runaway bloom of a microorganism similar to the red tide and other periodic blooms that pollute today's rivers and bays. The microorganism called *Methanosarcina* is believed to have rap-idly spread globally through the seas at that time. These creatures like carbon dioxide and have the ability to manufacture methane, and they were certainly doing so near the end of the Permian period. All this extra methane could explain the rising tempera-tures and the deposition of that unique carbon in ocean sediment. If this occurred in conjunction with the massive eruptions of the Siberian Traps at that time, the double whammy might have been too much for our planet to handle.

Incidentally, those Siberian eruptions probably didn't initially heat the world. They spread ash over Asia, buried an amazing 770,000 square miles in lava, and obliterated entire forests via the fires started by the lava, and this production of ash and smoke may have initially cooled Earth, so much so that ice formed, sea levels fell, and the later carbon-induced temperature increase had an easier time wreaking havoc on the planet in its new fragile state. In any event, we should never underestimate the malignant potential of prolonged volcanism. In 1783 a volcano called Laki erupted in Iceland. As one paleontologist explained, "Within a year global temperature dropped almost two degrees. Imagine a Laki erupting every year for hundreds of thousands of years."

In any case, we've seen enough to understand why, these days, researchers remain divided as to the P-T extinction's cause. In addition to the sudden events just cited—and some evidence indeed indicates that the extinctions happened in a "brief" two-hundred-thousand-year period—other evidence shows that the world's biosphere had already been slowly suffering. For example, only two trilobite species remained to go extinct; the others had died off in the previous ten million years.

In the forefront of the "slow" theories, we also find, of all things, the supercontinent Pangaea, a single enormous landmass that later broke up into today's familiar continents. The final formation of Pangaea happened just before the P-T extinction and certainly had been radically changing global air and water currents, causing a dramatic worldwide cooling event.

So there you have it, the greatest cataclysm in our planet's living history. Ninety percent of all species vanished. Creatures like the lovable trilobites that had been around since forever vanished. More than 95 percent of the things in the oceans died. So did the forests and the insects. Not even one in three land species survived. And recovery was interminably delayed; it was just too hot.

And we can only guess why it happened.

CHAPTER 19

THE DINO SAUR SHOW
GETS CANCELED

Comets, as we've seen, have been feared throughout all of recorded history, even if they're actually almost always harmless. But rarely—never during our human occupation of Earth but three times previously—has a comet or asteroid actually slammed into our planet. The results were always cataclysmic.

The surest, most detailed apocalyptic impact is also the most famous because it explains the sudden extinction of the dinosaurs. It happened sixty-six million years ago, and its discovery makes for a cool science story, especially since the man responsible was also instrumental in major planetary violence of a different type. His business card didn't say LUIS ALVAREZ, CATACLYSM CREATOR. But perhaps it should have.

Luis Alvarez was born in San Francisco in 1911 and earned his PhD at the University of Chicago, returning us again to our Windy City motif. After working for Ernest Lawrence in the Radiation Laboratory at the University of California, Berkeley, Alvarez joined the Manhattan Project and helped create the first atomic bombs. He was especially instrumental in developing the weapon's detonators, the high explosives and the lensing system that focused the blast of conventional explosives inward so that it evenly compressed a fifteen-pound ball of plutonium from all directions simultaneously, turning it instantly supercritical simply by shrinking that soft metal sphere from the width of a softball to that of a tennis ball. He got to witness the results firsthand, as he was present at the initial Trinity A-bomb test in 1945 and was an observer on one of the three planes that took part in the Hiroshima bombing.

A decade or so earlier, he was one of the discoverers of the strange east/west effect, the finding that cosmic rays arriving from deep space have preferred directions of approaching our planet, thanks to the interplay of their electrical charges and Earth's magnetic field. Since cosmic rays are created by violent events such as supernovas, it is clear that Dr. Alvarez was already intimately involved with the consequences of cataclysmic occurrences when he helped create some more, in the form of atomic weapons. This is the background against which he and his son, the geologist Walter Alvarez, made the discovery for which they will forever be best known.

After winning the Nobel Prize in Physics in 1968, Alvarez focused on the strange sudden extinction of the dinosaurs. Researchers had already suggested in 1953 that perhaps a cataclysmic impact event — a comet or asteroid smashing into Earth — was responsible for the radical change on Earth at the end of the Cretaceous period. But this remained merely an intriguing idea until Walter and Luis devoted over a year to uncovering inarguable evidence of the event.

Back in 1980 when the Alvarez paper was published,[1] the concept of mass extinctions seemed more like sci-fi or pop science. It was novel and controversial. Having laboriously moved away from Earth's biblical-type sudden-occurrence scenarios that had dominated human thought for over two thousand years, researchers were quite happy to visualize global changes as being languid events that unfolded over millions of years. They had numerous areas of solid evidence to support such lethargy.

Climate gradually altered as tectonic plates collided and new mountain ranges arose, each of which influenced weather patterns, upper winds, sea levels, carbon dioxide concentrations, and hundreds of other interrelated factors whose relations were complex even for the latest ENIAC punch-card computers tackling them. Serbian theorist Milutin Milankovitch showed that Earth's regional receipt of sunlight varied over separate cycles of 23,000 years (due to changes in how the Earth wobbles on its axis), 41,000 years (thanks to changes in its axial tilt), and 100,000 years (due to alterations in the shape of its orbit around the sun), so clear episodic mechanisms for ice ages and other gradual planetary alterations were firmly in place. Obviously, earthly life would be affected by these. Probably some ice age had finished off the dinosaurs.

But there was a problem, and it strongly suggested a bizarre abrupt event of unimaginable violence. The fossil record showed that all the dinosaurs everywhere had died suddenly. They weren't just peacefully dropping dead from high cholesterol. One moment they were all over the planet after having lived and evolved for tens of millions of years, and then, in a matter of a few years, they were gone, poof.

Luis's son, Walter, was an obsessive geologist. He knew that the disappearance of the dinosaurs and many other simultaneous changes in Earth's biosphere had happened at a geological time marked by an abrupt alteration that could be dramatically seen in the Earth. It was a layer, a remarkable boundary, at which the color of the rocks and earth suddenly changed from dark to light. You could view it in countless cliffs and highway cuts that were obvious from roadside rest stops.

So the Alvarezes went around the world and analyzed the composition of this sediment layer to see if anything noteworthy had been deposited there during this moment in history when the dinosaurs had vanished.

They got a shock. No matter where in the world they looked, this boundary marking the transition between the Cretaceous and Tertiary periods (the latter now called the Paleogenic)[2] was enriched with the metal iridium.

Iridium is extremely rare on Earth. But it's richly found on meteorites, almost all of which are fragments of asteroids. Iridium is also seen in enhanced amounts on comets. The silvery metal is thus an extraterrestrial fingerprint. Apparently, dusty meteoritic debris had filled the air sixty-six million years ago (the Alvarezes' original paper gave the time as sixty-five million years ago), and that dust then settled down to leave its imprint in the fossil record everywhere.

From that discovery, the conclusion was both obvious and amazing. Earth had been clobbered by a giant meteor—an asteroid or comet—whose detritus had filled the atmosphere. From the amount of material required to create the iridium layer, the Alvarez father and son calculated the volume of the interloper. The celestial body that delivered this disaster would have had to be at least six miles across, or about the size of Manhattan Island, which meant a bit larger than Mount Everest. For Earth's biosphere, it had been a true cataclysm.

Scientists are normally skeptical to the point of crankiness. You don't go overnight from believing in an achingly slow-changing Earth to an acceptance of such a totally new paradigm. A mass extinction? The disappearance of 75 percent of the world's plants and animals? The death of every dinosaur in a couple of years? And then a sudden opening of a new biological niche that was promptly filled by small warm-blooded creatures like rats? Which then evolved rapidly until, before you knew it, in a few tens of millions of years, here humans were with their VCR players.

It was the fastest revision of scientists' view of Earth's timeline in the modern era, and its implications ranked right up there with Darwin's *On the Origin of Species,* but in reverse, since Darwin had greatly slowed down the assumed timeline of events while the Alvarez family was speeding it up. It was probably the only model reversal ever to gain instant wide acceptance. But the Alvarezes had earned it. They hadn't merely created a logical hypothesis; they had reams of data painstakingly culled from C-T layers around the world, complete with laboratory-determined iridium-layer quantities. They'd found that the thin clay layer representing the C-Pg boundary had an iridium density of over six parts per billion, which was fifteen times more concentrated than Earth's average crustal iridium inventory. So they had made their case ironclad before they published their 1980 paper.

In the nearly forty years that have passed, the evidence for the cataclysmic mass extinction has only grown stronger. A 2016 reanalysis fine-tuned the time of the explosive event from sixty-five million years ago to sixty-six million, with an uncertainty of just eleven thousand years. Even better, the impact location was found.

It's not known whether the popular 1940s and 1950s star who was born Frances Rose Shore derived her stage name from the suspiciously similar *dino-saur*, which means "terrible lizard." *(Paramount Pictures)*

Such a violent collision between an asteroid and Earth could not fail to leave a scar. Unknown to Luis and Walter Alvarez, it had already been seen by two geophysicists looking for petroleum deposits in the late 1970s, a few years before the dinosaur-extinction paper was published. In 1991, others positively identified it as the impact location.

In the eastern Yucatán Peninsula of Mexico, essentially off-shore, the giant 110-mile-wide, 12-mile-deep crater sits near the town of Chicxulub.[3] In 2016, scientists laboriously drilled a mile into the seafloor just off the Yucatán coast to obtain rock-core samples from the peak ring of the crater. Radiocarbon and other dating methods showed the material included nonterrestrial components that arrived 66,038,000 years ago.

Now, armed with specific data about the impact size, we can be confident about the exact violence that was inflicted on our world. The asteroid, already estimated to be six or seven miles wide, has been upgraded to nine miles wide, although some think it could be as much as seven times bigger than that.

We can visualize the cataclysm unfolding. First the incoming asteroid, larger than Mount Everest. Though it was traveling at around twenty miles a second, or seventy thousand miles an hour, it would have *seemed* fairly slow as it approached. That's because — as this author repeatedly demonstrated in his 2014 book *Zoom*[4] — an object or animal's perceived velocity critically depends on *whether the moving object is able to rapidly displace its own diameter.* A horse sauntering at three miles an hour will walk its own length in two or three seconds and will not seem particularly fast. But a mouse moving at that same speed will traverse fifteen of its body lengths per second. Displacing its form fifteen times over per second makes it seem supersonic. Conversely, a jumbo jetliner approaching for a landing often appears to be virtually hovering in place even though it's traveling fifty times faster than the scurrying mouse.

This is why a shooting star seems superfast. A meteor is a mere bright dot in the night sky. (Actually, the typical visible meteor is indeed only the size of an apple seed, although the globe of glow-ing, ionized air surrounding it is a few inches wide.)

The meteor is therefore traversing thousands of its own diame-ters per second and thus appears superfast. But on that day (or night; we don't know which it was) sixty-six million years ago, the giant meteor, at least six miles across but possibly as wide as sixty miles, was scarcely moving its own width per second. That's about twenty times slower-seeming than a bowling ball heading toward the pins. Not exactly lethargic, but not brisk either.

No matter. When it hit the Earth, traveling more than forty times the speed of sound, its impact unleashed the same energy as ten billion Hiroshima bombs. Giant tsunamis the height of sixty-story buildings spread across the Caribbean. Violent shock waves and insane earthquakes of magnitude 10 distressed the ground, causing massive die-offs of numerous ocean organisms five thousand miles away. Meanwhile, uncountable particles of various sizes, the detritus from the impactor and from pulverized sections of Earth's crust, were hurled skyward and then arced downward, all the fragments moving as fast as meteors. They were incandes-cently hot, and their sheer numbers heated the air like a furnace as they descended and started fires in forests everywhere.

Smaller particles—soot—combined with water were thrown skyward; they darkened the air and shut down photosynthesis for years. Tons of rocks buried entire countries. Everything dependent on plants died within a week or two. Animals dependent on vege-tarian creatures for their own sustenance died soon after. The dino-saurs were gone.

It should be mentioned that, for those of us who would like to realistically visualize the events of that day, researchers continue to evaluate the data based on the worldwide iridium concentration at hundreds of locations in the C-Pg layer, the Chicxulub crater size and other characteristics, and the core sample rocks under that area in an attempt to obtain ever more accurate estimates of the energy released in the cataclysm. The most current figures indicate a blast in the area of 10^{24} joules and perhaps as high as 6×10^{25}. After the cataclysm, the oceans required three million years to return to normal.

Even the celestial nature of the impactor is being revisited. Walter Alvarez believed it was a stony-type asteroid, but some are arguing, based on the iridium concentration and impact speed, that it was a long-period comet.[5] Yet another consequence was the instigation of searches for previous cataclysmic collision events, and investigations to determine if there was some sort of periodicity to such disasters, which would suggest that perhaps our sun is a member of a double-star system whose companion periodically ventures near enough to gravitationally disturb the countless small bodies of the Oort cloud or Kuiper Belt, changing their orbits enough to hurl them toward the inner solar system. If so, a further encounter with giant Jupiter might provoke just the kind of perilous orbital change that would send the objects in our direction.

An early apparent periodicity in such collisions—complete with a dim red theoretical companion star (named Nemesis)—has morphed in recent years to a conviction that no such binary system exists. The collisions are indeed random. And so, too, regardless of their origins, are the mass extinctions that have plagued Earth with the most terrifying of all cataclysms.

CHAPTER 20

SNOWBALL EARTH

What's the worst imaginable global situation? In our current era when we worry about global warming, the concept of ice might not seem so terrible. But now picture the entire planet covered with glaciers from pole to pole. That's the Snowball Earth scenario. A fair amount of evidence suggests that this actually happened about seven hundred million years ago.

We already know about periodic ice ages, which were instigated, in all likelihood, by the Milankovitch cycles discussed elsewhere. But imagine during one of these cold eras caused by changes in our planet's axial tilt and orbital shape that some further instigating factor arises, such as a series of super-volcanoes that block sunlight for decades. Now the burgeoning glaciers continue to spread southward, perhaps prodded by a rare alignment of sun-screening orbital conditions configured in the most ice-promoting orientation. Studies reveal that if glaciers ever get to about twenty or thirty degrees from the equator, no natural process will stop them from continuing to the equator itself. Then we're a white planet. In this situation, the global mean temperature would stabilize at about –74 degrees Fahrenheit, and even the equator would be only about –10 degrees.

Worse, such a situation locks itself in. That's because snow and ice have a very high albedo, or reflectivity. Liquid water reflects only about 10 percent of sunlight, which means 90 percent is absorbed and makes the water warmer; it's why the oceans appear dark. Land-covered regions with typical vegetation reflect 30 percent of the sunlight hitting them. But ice reflects about 50 percent of the sunlight, and fresh snow reflects around 90 percent. This means that a snow-covered area can barely get warmer even in full sunlight, so once a completely ice-covered Earth has formed, its high albedo creates a positive feedback loop that maintains the condition.

This is why many scientists were initially skeptical about the Snowball Earth idea. They said that once such a situation developed, it would remain forever because no amount of sunlight would melt it. And since that condition obviously doesn't exist today, it means it could never have happened.

Indeed, researchers have established between three and five separate global conditions that are stubbornly stable. One of them is an earthly situation termed the Snowball, which automatically arises and maintains itself if the solar energy we receive diminishes by more than 1.6 percent. Needless to say, this would be a cataclysm. There would be no exposed land on which plants could grow, so no animals could survive. Earth's surface would become sterile, although the undersea world, though threatened, might hang on.

The ice would thicken over time. That's because even with the cold and dry climate, the air would still transport water vapor from areas where the ice had sublimated (meaning it changed directly from a solid to a vapor).[1] This atmospheric vapor ultimately produces additional snowfall that thickens high-latitude glaciers.

If this disastrous scenario really happened, the flowing glaciers would rip up the underlying ground and deposit telltale signs of their passage, such as drop stones, tills, moraines, eskers, and glacial erratics, that would broadcast the glaciers' existence even half a billion years after they were gone. Seeing such deposits in equatorial regions has convinced many researchers that the Snowball Earth cataclysm really happened.

The evidence for this comes from two separate time periods. The first episode, about 2.3 billion years ago, may have been triggered by the sudden oxygen cataclysm described in chapter 18. But the Snowball Earth event that has garnered the most attention was the one about seven hundred million years ago.

The global-glaciation idea was first proposed by the Australian geologist and Antarctic explorer Douglas Mawson and supported in a 1964 paper by W. Brian Harland, who offered paleomagnetic data claiming to prove that glacial tillites in Greenland were also deposited in the tropics. The 1960s also brought energy-balance climate models that showed that once the process was initiated, the entire Earth would indeed be covered in ice and that this new condition would remain in a stable equilibrium.

An ice-covered Earth is, unfortunately, one of the five possible stubbornly stable conditions in which our planet tends to establish a long-term equilibrium. (Discover *magazine*)

The catchy term *Snowball Earth* first appeared in a 1992 paper published by Joseph Kirschvink. He also offered further evidence for it, such as the presence of banded-iron formations and the first realistic escape mechanism that could explain how our planet emerged from such a dreadful state. Then, in 1998, a persuasive paper in the journal *Science* by Paul F. Hoffman used the presence of cap carbonates to support the Snowball Earth hypothesis and offered evidence that the period was quickly followed by a change to unusually hot global conditions.

Well, if this really happened, it was a cataclysm, no two ways about it. But you know scientists; if they can shoot something down, they will. These days, Snowball Earth is controversial, although the scientific consensus seems to be swinging in its favor. One of the ongoing problems in accurately determining conditions three-quarters of a billion years ago is the shift of tectonic plates over time. When sedimentary rocks first form and cool, their magnetic components align themselves with Earth's magnetic field as soon as they drop below the Curie temperature of 1,414 degrees Fahrenheit, and this will theoretically tell us where the rocks originated. Those that came from near the equator but contain glacial sediments such as drop stones certainly suggest that ice covered the tropics.

But what if the magnetic poles were in a different position way back then? Subtracting all the ambiguities that researchers have fought over leaves us with only one sure equatorial glacial layer — the Elatina deposit, which has tectonically migrated to Australia.

Other supportive evidence includes tropical glacial deposits as much as sixteen thousand feet thick separated by bands of nonglacial sediments, the latter bands suggesting that glaciers repeatedly melted and reformed over millions of years, since it's hard to envision an origin for such deposition layers under continually frozen oceans.

Moreover, if Snowball Earth really happened, the constant stream of disintegrating micro-meteors that are still estimated to add forty thousand tons of mass to Earth annually would have caused a sprinkle of iridium to be laid down on the snow, ice, and glaciers. Then, when the glaciers disappeared, we would find this odd iridium deposition layer on seabeds everywhere, which would serve as evidence for the ice-covered condition. This is indeed observed. But does this prove the Snowball Earth scenario? Probably, but maybe not, since periodic asteroid impacts would also create enhanced iridium layers.

In a January 2000 *Scientific American* article, Paul Hoffman offered evidence showing how cap carbonates are found side by side with ancient glacial deposits at places that were once very near the equator. As he explained, "Thick sequences of carbonate rocks are the expected consequence of the extreme greenhouse conditions unique to the transient aftermath of a snowball earth. If the earth froze over, an ultrahigh carbon dioxide atmosphere would be needed to raise temperatures to the melting point at the equator. Once melting begins, low-albedo seawater replaces high-albedo ice and the runaway freeze is reversed." In other words, the same volcanically produced high CO_2 that was sufficient to save the planet from its snowball state did not magically vanish once the ice was gone. Instead, it lingered to produce one of the hottest periods on record.

The genetic assaults that Snowball Earth imposed on Earth's microscopic life (and it was *all* microscopic up to that point) and the newly oxygenated atmosphere explored in chapter 18 were probably the triggers for the Cambrian explosion that followed soon after. This was the dramatic sudden appearance of multicellular life throughout the planet, the results of which are seen everywhere we look.

Though the evidence for Snowball Earth is good, skeptics enjoy a compromise idea called Slushball Earth, proposed by American geologist Richard Cowen. He suggested that when the massive ice sheets covered much of Earth, the ocean areas around the equator were wrapped only in a thin, slushy water-and-ice mixture that probably included areas of open sea. This would have allowed plant life in the ice-free ocean regions to continue to photosynthesize, letting them survive the era when the rest of the planet lay beneath snow and ice.

Whether Slushball or Snowball, the catastrophic era lasting ten million years probably happened, and it was followed by a time of superhigh volcanic activity whose resulting extreme CO_2 levels produced equally high sea levels, since all that ice had been turned into seawater.

Researchers have now shown that an extreme and extended period of volcanic activity, with detritus blasting through the white surface, would indeed have sufficiently increased Earth's carbon dioxide to have finally allowed the ice to melt.

So, as unpleasant and seemingly endless as it was, it too passed.

CHAPTER 21

THE PLAGUE

Nearly all our cataclysms, celestial and terrestrial, revolve around physical events. But occurrences seven centuries ago force us to head in a very different direction and include two biological catastrophes that meet all the usual Big Bummer criteria. The first, in the mid-fourteenth century, was shocking and deadly enough to be labeled, in a 2005 *History Today* account, "the greatest catastrophe ever."

We're talking about the famous Black Death. Imagine the terror of a single malady engulfing most of a local population and killing more than half its victims in a matter of days. And that's ignoring people's responses to the pandemic, which in many places were as brutal as the disease itself.

When the Black Death first struck, in 1346, it ravaged Europe for seven interminable years and then returned unpredictably over and over again for the next three centuries. It killed more than half the population, with women and children succumbing at a higher percentage than adult males. Nobody had a clue what was causing it or how to avoid contracting it. And while human-borne diseases like tuberculosis are naturally concentrated in places of greatest population density—the cities—this scourge struck urban and rural areas alike. Still, the medieval cities, in the best of times, were crowded with animals, feces, open sewage, and a constant stench and filth, so the misery of the suddenly agonized and terrified city dwellers added but a finishing horrendous touch to an unpleasant preexisting stage. Few fled, because word had spread that the farmlands were being ravaged with at least equal ferocity.

The real centers of plague were the ports. Venice was slaughtered, and London lived with wave after wave of the disease. All Mediterranean and then northern European ports were epicenters. Not a single one was exempted.

All social classes were affected by the Black Death, although the elite had the financial ability to flee to distant places. Isaac Newton received his degree from Trinity College in 1665, just as the bubonic plague was descending on London, and he fled to his family's farm and didn't return until the plague had passed. Yet those who fled often took the plague with them and spread it to the countryside. In the end, there was no refuge.

What was it like?

A letter writer in Florence said this:

All the citizens did little else except to carry dead bodies to be buried [and] at every church they dug deep pits down to the water-table; and thus those who were poor who died during the night were bundled up quickly and thrown into the pit. In the morning when a large number of bodies were found in the pit, they took some earth and shoveled it down on top of them; and later others were placed on top of them and then another layer of earth, just as one makes lasagna with layers of pasta and cheese.

Well, granted, the culinary metaphor doesn't quite work, and neither does any other smooth transition to the cause of it all — the black rat, also called the house rat or the ship rat. It was the plague that gave this highly intelligent mammal the dreadful reputation it would never shed.[1]

The story would be quite different if only the fleas that carried the bacteria *Yersinia pestis* had been attracted to that other rat, the brown rat, which is shy and secretive and avoids people and their homes in favor of sewers and such. Instead, yersinia infects fleas that glom on only to that black, sneak-around-your-kitchen-and-basement variety, a kind that often stows away on ships. If the rat colony is dense, the disease spreads widely in their midst. After around two weeks, the bubonic plague kills off virtually the entire rat colony, and then the fleas, increasingly hungry, grow ever more desperate and fidgety for the next two to three days before they finally start biting humans.

You'd think the townspeople would have noticed the dying or disappearing rats, put two and two together, and figured out the cause. But with all the European plague episodes (in 1374, 1400, 1438–1439, 1446–1453, 1456–1457, 1464–1466, 1481–1485, 1500–1503, 1518–1531, 1544–1548, 1563–1566, 1573–1588, 1592–1593, 1596–1599, 1602–1611, 1623–1640, 1644–1654, 1664–1667, and a few more epidemics in the eighteenth and nineteenth centuries), no one ever did.

After a contaminated flea bites a human, the pathogen travels to a lymph node that swells to form an alarmingly large bulb, usually in the groin, in the armpit, or on the neck. These bloated, painful buboes give the disease the name bubonic plague. *Yersinia pestis* produces toxins that paralyze the victim's immune system.

The infection takes half a week to incubate before the person feels ill, and then it's a fight for life with a precarious, doubtful outcome. If the disease spreads to the lungs, it's almost always fatal, and this also introduces an agonizing period of difficulty breathing. If the patient survives the next half a week, she may beat the disease, and even acquire a permanent resistance to future exposure, not to mention a roomful of sympathy cards. But in 80 percent of cases, the victim dies.[2]

The total time from when a region's rats are first exposed to infected fleas—usually from contact with fleas from rats on a merchant ship that has just docked—to when people begin to die is a remarkably reliable average of twenty-three days.

The black plague, like polio, exhibits an unfailing seasonal pattern favoring the summer and early autumn. This plague blueprint occurred everywhere without exception. For example, in Norway, which saw thirty waves of plague over several centuries, there was never a winter epidemic. Viral respiratory diseases favor winter, when people are closest together and when the airborne droplets most easily pass from person to person. But plague is an insect-borne bacterial malady. In most places, the cold months put the plague on hold until the return of warmth. In some communities, the winter totally ended the epidemic, and it did not reappear the following year.

Not all epidemics are as egalitarian as the Black Plague. In a few diseases, the affluent are preferentially struck. Polio, for example, can sometimes be a harmless virus for those who acquire it in infancy. But members of the middle and, especially, upper class, protected from crowds who might be harboring the virus, may first encounter the pathogen later in childhood or in early adulthood, when its effects are more likely to be horrific. We saw this not just in famous people such as Franklin Roosevelt but in heartbreaking images from the 1940s and early 1950s of polio victims in iron lungs. They mostly displayed the neatly manicured nails and well-groomed hair of the well-to-do.

But no stratum of medieval society was isolated from rats and fleas. And the disease was carried far and wide, not just by rat-infested trading ships that traveled an average of a hundred and eighty miles a week, but also by people. Although *Yersinia pestis* in human blood is five hundred times less concentrated than it is in rat blood, people blood was never the real problem. Rather, those arriving from a plague-stricken area often had clothes or luggage contaminated with diseased fleas. Then, twenty-three days after the travelers set foot in the new town, like clockwork, the region would see its first fatalities. The new local rat colony now carried the infected fleas.

The first recorded epidemic wasn't primarily a European scourge; it mostly attacked the Byzantine Empire in the middle of the sixth century and was commonly called the Plague of Justinian, after the emperor of Constantinople, Justinian I, who actually came down with the illness but survived. This earliest bubonic plague pandemic killed an estimated twenty-five million if one counts only that first outbreak; it was double that if one includes its recurrences through the next three hundred years. Then there was a pause. It was another five hundred years before the next bubonic plague pandemic, the one that came to be known as the Black Death.

It is the iteration of the bubonic plague known as the Black Death that captures our imagination. When Shakespeare wrote "a plague o' both your houses," he meant something terrifying, as you can tell when he has King Lear call his daughter Goneril "a plague sore, an embossed carbuncle in my corrupted blood." And there is probably not a better description of the horror of this pestilence than this: "The dead man's knell / Is there scarce asked for who, and good men's lives / Expire before the flowers in their caps."

In fifteenth-century Europe, the horrible fact that the plague had arrived was often noted in records by clergymen and town officials. Locals would become aware that the plague was breaking out and that a full epidemic had arrived after an average of about fifty days in towns and small cities. But this delay in sounding an alarm probably didn't much matter. There was nothing anyone could have done except, perhaps, try to get very far away.

As we've seen, the plague was maintained and propagated by trade, particularly by ships and the rats they carried. In many cases, the disease killed every rat on a given ship before the vessel arrived at its new port, but the fleas usually managed to survive until reaching land, whereupon they would promptly infect the local rat population to start the twenty-three-day timer ticking once again.

The Black Death killed around 60 percent of Europe's population. Given that the continent's census was around eighty million at the time, this means that some fifty million lives were lost, or about one out of five humans then alive on the planet.

For the sick and dying, the cause was almost always attributed to God being angry, and the standard treatment was to be penitent and perform rituals that would hopefully be acceptably appeasing. In some places, astrological causes were blamed, such as a conjunction of Jupiter and Saturn. Sadly, the scourge was also often blamed on those with skin rashes or even common acne, and innocent people were sometimes put to death as a preventive measure. Jews were scapegoated; in Strasbourg, Barcelona, Basel, and Flanders they were slaughtered.

According to a 2005 plague analysis by Ole Benedictow in *History Today,* the true underlying cause of this incredibly virulent slaughter was the leap in European population in the High Middle Ages, especially in the twelfth and thirteenth centuries. All the new people needed to be fed, and traditional agriculture could not accommodate this, which necessitated trade. The expansion of ports and commerce centers, with their growing number of vehicles and ships, ensured that "contagious diseases reached even the most remote and isolated hamlets."

The biggest trading centers suffered the most. Outbreaks of bubonic plague happened twenty-two times in Venice in the 170-year period starting in 1361. The later plague outbreak there, from 1576 to 1577, killed fifty thousand in Venice alone. Over 60 percent of Norway's population died from 1348 to 1350. Benedictow wrote, "In the first half of the 17th century, a plague claimed some 1.7 million victims in Italy, or about 14 percent of the population," and that "in 1656, the plague killed about half of Naples' 300,000 inhabitants."

Modern studies show that some countries suffered only a 20 percent plague mortality, while in others it was as high as 80 percent. Of course, the disease also ravaged the Muslim nations, and Russia, and indeed seems to have been endemic in the Crimean region, where it remained in wild gerbils and from where it would be periodically launched westward to Europe on new generations of fleas and rats.

The Black Death finally died out in Europe in the seventeenth century. It had come out of China along the Silk Road. The Silk Road is a two-way street, and DNA analysis has now shown that the outbreak that began in China in 1850 originated in Europe. In 1894, it reached Hong Kong, devastating the population. It is here that Alexandre Yersin discovered the bacillus that bears his name. And it was here that the rat and the flea and *Yersinia pestis* met the steamship, and this is when and how the bubonic plague was taken around the world, including North and South America, for the first time.

Recent DNA sequencing has produced some interesting controversies. Many but not all epidemiologists and disease experts feel that the strain of *Yersinia pestis* responsible for the original and most virulent recurring plague epidemics is now extinct and that, should there be any major outbreak in the future, it will have a much lower mortality. Already, modern antibiotics have reduced the fatality rate to around 11 percent even in third-world outbreaks, where treatment is not begun as optimally early as might be desired.

Thus we have real hope that the bubonic plague as a cataclysm is now, in terms of massive deaths, mostly one for the history books. But not, sadly, entirely. As I write this chapter in 2018, there is an outbreak of plague in Madagascar. The death toll so far is two thousand.

Benedictow, in his 2005 article, included a contemporaneous account meant to illustrate how utterly deep was the fear and suffering, in this case during the 1346 to 1353 outbreak in Florence. It was written by the renowned Renaissance poet Petrarch, who sends a brief touching message to us through time:

O happy posterity, who will not experience such abysmal woe and will look upon our testimony as a fable.

CHAPTER 22

JUST THE FLU

Placing our medical chapters side by side is no accident. The grouping arises chronologically despite the fact that the plague seems to be an event in the distant past. In truth, plague epidemics have continued right into modern times.

This section of our cataclysms review, dealing solely with earthly calamities, took us to the sixteenth century in the previous chapter. We then quickly passed over the seventeenth to nineteenth centuries, which witnessed no cataclysms, or at least none that met the gruesome standard of thirty million fatalities. But the twentieth century, sad to say, had no fewer than three events that each killed over 1 percent of the world's population, which is another one of our cataclysm benchmarks. This is why we now logically jump to 1918.

We're not talking about World War I, which wasn't chopped liver but nonetheless killed "only" sixteen million. Instead, we're exploring the familiar H1N1 virus during the single time when it behaved very, very strangely.

Plagues and epidemics have always been scourges, and one might argue that Russia's horrible mid-nineteenth-century cholera epidemic deserves a chapter. And also the Asian flu outbreak of 1957 to 1958. And the 1968 Hong Kong flu. Each of these did indeed take one to two million lives, making them true tragedies. But to cross the blurry line from scourge to cataclysm, we've, again, adopted that extremely high casualty cutoff, requiring an astonishing death toll exceeding thirty million. This has unambiguously happened only two or three times in our planet's disease history. The first time was the plague. The second began just as the Great War was winding down and troops were preparing to return home.

This deadly event generated surprisingly few headlines at the time. The pandemic's casualties blurred together in the public mind with those who never returned from the great European bloodbath. The war's deaths carried a far more dramatic impact than the flu's, especially for the news media, which perhaps explains why Ernest Hemingway and F. Scott Fitzgerald, writing bestselling novels in the years immediately following the Spanish flu pandemic, never mentioned the disease once. Moreover, the pandemic lacked a punchy name.

When it started being called the Spanish flu, to Americans, the label sounded both distant and trivial. Plus it was a misnomer, since it had nothing to do with Spain; the disease neither originated nor was particularly virulent there. The name arose because Spain's neutrality in the Great War kept its newspapers free, and they alone reported on this new disease menace; other European countries were censoring all reports of medical losses for fear of harming morale. Since it was solely headlined in Spanish newspapers, it got to be called the Spanish flu.

The later name, the 1918 flu pandemic, barely sounded any more malignant. But to those who were there, it was both horrible and strange.

We now know that the illness that first showed itself in the autumn of 1917, started sickening enough people to attract attention early in 1918, and ultimately killed fifty million before ending abruptly in 1919 was indeed influenza. It was a virus that nowadays we call H1N1, initials that stand for the glycoproteins (proteins with sugar-like side chains) hemagglutinin and neuraminidase. These are tools the virus uses to sicken humans, pigs, or birds, the three species most susceptible to its attack. Hemagglutinin lets the virus attach itself to animal cells, while neuraminidase is an enzyme that helps the virus reproduce. Strains of H1N1 are considered influenza A viruses, and even in modern times they recur in many winter influenza outbreaks. They are widespread in swine and birds as well, and this lethal Great War strain probably arose one day in Kansas when bird and human viral genetic material was exchanged, possible in a single fateful swine infection.

The 1918 flu outbreak was bizarre. First, it infected half a billion people around the world, which was about a third of the global human population. Yet what was particularly strange was its lethality. A normal flu epidemic kills about one in a thousand people who contract the disease. But the 1918 Spanish flu killed one out of every ten of those sickened in vulnerable age groups. Speaking of which, the age of those who died was even more peculiar. A normal influenza outbreak kills the fragile and vulnerable, with the most deaths among the very young and very old. But the Spanish flu had a prominent lethality in what is normally the hardiest age group, the twenty- to forty-year-olds.

Even the pattern of illness was unprecedented. A wave of flu was initially seen at the beginning of 1918. The location of the very first case has been hotly debated for a full century now, with some evidence for a start in January in Haskell County, Kansas, where a local physician was alarmed enough to report the sudden outbreak to the U.S. government, an extraordinarily unusual thing to do at that time. What's indisputable is that on March 4, 1918, at the Camp Funston army military base in Fort Riley, Kansas, the company's cook, Albert Gitchell, came down with the flu. Exactly one week later, one hundred other soldiers were in the base hospital. The casualty toll there skyrocketed to 522 men by midmonth. At this same time, flu was breaking out in the New York City borough of Queens. At that point, nothing seemed unusual, since influenza outbreaks had been familiar nuisances every few years in the previous several decades, although there hadn't been a particularly deadly outbreak for thirty years.

Everything changed in August. Suddenly a new wave of sickness appeared in far-flung places simultaneously, and now it had a much higher mortality. The very first cases were seen in three busy port cities that were involved with war shipments and troop movements: the French military port of Brest; Boston; and the African capital of Freetown, Sierra Leone. It quickly became obvious that something very weird was happening.

Although the symptoms were flulike, around 20 percent of the victims developed especially acute respiratory problems and could not be saved. At the author's request, the strange medical situation was analyzed exactly one century later by British infectious disease specialist Dr. John Froude. He held a definite opinion about a cause of the victims' 1918 lung distress, and he shared it in an e-mail to the author: "The virus has a tropism (physical attraction for) the alveolus, the little sac at the end of the bronchi where oxygen exchange takes place. It fills up with protein and you cannot oxygenate. This is called ARDS, or acute respiratory distress syndrome. When they dug up Lucy in Alaska and looked at her permafrost-preserved lungs, this is what they saw."

The illness spread quickly and killed about 5 percent of the planet's human population. Oddly, 99 percent of those who died were under the age of sixty-five — the opposite of what was normally seen in flu outbreaks or what had been observed in the first wave of illness early that same year.

The big stories were not what places influenza was ravaging but the places where it was absent. People on ships were carrying the illness, and isolated communities came down with the disease soon after boat dockings, a pattern that resembled that of the plague. In German Samoa, nearly a quarter of the population died of the Spanish flu in 1918. But after Governor John Poyer imposed an early blockade around American Samoa, not a single case of the illness was reported there.

Yet another oddity was the season that it struck. Normally a flu epidemic hits in winter, when people are most crowded indoors and when the heated air is driest, affording viruses their best chance to invade nasal passages. And while the first wave of the 1918 flu did indeed strike in midwinter, the far deadlier outbreak swept the globe in summer and into autumn. This was obviously a mutation of the earlier virus, since those who recovered from the milder initial infection were immune to the second wave. It was still at full force in October. But by mid-November, it was completely over, and no new cases appeared, although after months of inactivity, in the spring of 1919, a third, less deadly wave went around the globe. Apparently, the H1N1 virus had mutated to a less lethal strain. But before this cessation, the pandemic killed as many people from March 1918 to March 1919 as the plague had killed in a full century.

What had happened? Why had an ordinary, common H1N1 virus become so deadly? And why the bizarre pattern of lethality among young adults? While some mysteries remain, the explanations for this cataclysm have arrived in the past two decades. The odd lethality apparently had two main causes.

The exhumation of 1918 victims, especially of that Alaska woman mentioned by Dr. John Froude, the one whose body was preserved in permafrost, has let medical researchers examine the responsible virus. They found that it resembled viruses that had caused less deadly flu outbreaks in the latter half of the nineteenth century. Turned out, the adults who died in such great numbers in 1918 were too young to have been exposed to the earlier, similar H1N1 virus, and thus they were particularly vulnerable. Those who were over fifty-five years old in 1918 had been exposed as kids to influenza strains closely related to the 1918 flu, which gave them a significant degree of immunity.

But a totally different factor may have been primarily responsible for the lethality. The 1918 H1N1 strain had the ability to cause a severe autoimmune response in humans. This so-called cytokine storm is an extreme immune response in which the body overreacts to the pathogen, triggering excess white blood cells, resulting in often-fatal effects on the lungs. The patient sometimes hemorrhages from the nose, ears, and lungs—symptoms that accurately describe what happened to the victims of the Spanish flu. This would also explain the odd 1918 age-specific mortality, since young adults generally have the strongest immune responses. Old and very young patients, by contrast, have the weakest immune reactions, which in this rare instance would have protected them from the worst consequences of the pandemic. As John Froude told the author: "The immune system is imperfect and has a dark side. An infection, after all, is the interaction between the pathogen and your immune system."

But whether or not we now fully understand its odd lethality, the fact remains that the 1918 pandemic killed more people in a year than the HIV scourge has killed in the past forty. In the United States alone, 670,000 people died of the Spanish flu.

Nor was this the end of recent medical cataclysms. Smallpox proved to be the biggest killer of all time—by far—with an estimated three hundred million deaths just in the twentieth century, even though it was finally eradicated in the 1960s. AIDS is also way up there on the twentieth-century list of disease mayhem, with about forty million dead and forty million more living with the illness. "If these are added to the 1918 Spanish flu plus just the third Bubonic pandemic plague which continued well beyond 1900," Dr. Froude notes, "we see that the twentieth century was pretty cataclysmic from plagues alone."

CHAPTER 23

THE SECOND WORLD WAR

I hesitate to include war in this investigation of cataclysms. Partly it's an aversion to the notion that our species is capable of inflicting such calamity on itself. Another reason is the sad open-endedness of such inclusion. Which conflicts should be incorporated and which omitted? But in the end, I have already employed the thirty-million-deaths criterion in deciding how to pare down the countless historical disease pandemics, so I might as well remain consistent and apply that same standard to wars.

This strategy will create the positive upshot of limiting the recitation to a single conflagration — World War II, which by the average of authoritative accounts produced thirty-seven million deaths. This amounted to roughly 2 percent of the world's population.

Some might imagine that earlier wars were worse, but none came close, except, perhaps, several historical Chinese conflicts. Since those produced still-disputed casualty figures, I can hedge my bets by imposing an additional criterion and noting that I will confine this examination to wars that involved large parts of the globe, as opposed to local or regional conflagrations.

By any criteria, only two wars resulted in over thirty million deaths: the Second World War and the Mongol conquests of 1206 to 1368. Of those, only the former had a truly global scale.

Of course, one might just as easily deem that twenty million rather than thirty million deaths constitute a cataclysm, especially in earlier historical eras when the world had far fewer people. Thus if we include wars that, like World War II, killed 2 percent of the humans on Earth, we'd have to use the twenty-million death toll as our standard for all conflicts prior to the twentieth century. In that case, we'd then include three separate wars within China. Specifically these are the half-century Qing conquest of the Ming that began in 1618, the Taiping Civil War that began in 1850, and the An Lushan Rebellion that started in 755 and ran for eight years. Each claimed more than twenty million lives, which were astonishing death tolls for those periods of history.

We'd also have to include the brutal 1519 to 1632 Spanish conquest in Mexico that destroyed the Aztec Empire. And the wars between the waning Roman Empire and the Germanic tribes of Europe, whose brutal hand-to-hand combat raged from 113 to 596 and probably killed around twenty million people, although accurate casualty figures are not available. Some would also list the Great War, or World War I, which a few authorities cite as claiming twenty million lives, though a more accepted mean is closer to thirteen million.

And there you have all the wars that might be considered cataclysms. Many American Civil War historians might argue for its inclusion, given that the fighting was so intense, bullets from opposing sides occasionally collided in midair and fused together, but in terms of sheer death toll, it claimed "only" about eight hundred thousand lives.

Why was World War II so much more deadly than all the others? Beyond the global reach of the conflict was the brutality of modern airpower, where a single firebombing raid on a city like Tokyo or Dresden would kill eighty thousand people overnight. But also, separate wars were layered within that larger one, like the horrible so-called Second Sino-Japanese War, which included major Japanese atrocities against the invaded population; the Japanese killed an estimated twenty-two million Chinese civilians.

The Second World War casualty total also includes concentration-camp atrocities, in which entire segments of European society, such as Jews, Gypsies, homosexuals, Polish Catholics, and the disabled, were subjected to genocide. It would be hard, then, to complete this book on cataclysms with an exclusion of that war.

However, unlike all of our other chapters, where the causes and chronologies of specific catastrophes are examined in all pertinent facets, in this chapter, I omit detailing the complex and lengthy origins, developments, and concluding factors of the Second World War, mostly because many scholarly historical works are widely available and, moreover, because I have nothing new to add to those narratives.

Instead, let its mere citation be included for the sake of acknowledgment that this single complex human atrocity that unfolded from 1939 to 1945 does indeed sadly belong in any list of the cataclysms that have befallen our universe.

CHAPTER 24

NUCLEAR CATACLYSMS

When people think of cataclysms reported by the media in their lifetimes, few would omit three accidents afflicting nuclear power plants. Three Mile Island is listed as a disaster in the index of the 2017 *World Almanac,* despite the fact that not a single person was harmed by that 1979 core meltdown, nor was there any private property damage. Far more unambiguously deserving of our cataclysm trophy is the fearsome 1986 Chernobyl event, which quickly killed forty-nine plant workers and early responders and is likely to eventually claim somewhere between four thousand and ten thousand cancer victims.

More recently, the vast majority would probably deem Japan's 2011 Fukushima power plant accident a cataclysm, although it resulted in no radiation-induced deaths and the radiation exposure for people living in Fukushima was so small compared to background radiation that it may never be possible to find statistically significant evidence of cancer increases. One can reach half a dozen fatalities if one includes those who died during evacuation procedures or who died due to exacerbations of their medical conditions when they were forced to move from local hospitals.[1]

In the 2011 event sequence, it's even harder to apply the cataclysm label to the Fukushima nuclear mishap when it's compared to the far greater casualties caused by the precipitating earthquake and tsunami earlier that same day, or when it's compared to the Banda Aceh tsunami seven years earlier, which claimed 230,000 lives.

Still, two of these three nuclear events involved explosions. All had near-light-speed runaway chain reactions. If rapid motion is necessary to meet our criteria for a cataclysm, such frenzy was abundantly present each time. Besides, they all possessed fear factors that were off the scale, which caused them to be widely perceived as cataclysms. But more than that, they decisively shaped global opinion in a way that effectively brought an end to widespread nuclear power plant construction, and with that went realistic hopes for an end to carbon emissions. How deeply this will ultimately affect the planet's biosphere is not yet fully known.

When the twentieth century began, science was just realizing that the atom was not the smallest thing in nature, as had been believed since the ancient Greeks (who'd coined the word). The first subatomic particle, the electron, was theorized to exist in 1896 and discovered three years later. In the next three decades, physicists found the major components of every atom's nucleus. There was the proton, a heavy particle, the number of which in a nucleus is solely responsible for that atom being, say, oxygen as opposed to, say, carbon. Then there was the neutron, which was even a bit more massive. The neutron lurks in every element except ordinary hydrogen. It gloms on to protons and other neutrons in the nucleus, held there solely because of a truly strange attractive property physicists named, in an exuberant moment of poetic license, "the strong force."

Energy, lots of it, is released when an atom is altered, particularly if the strong force is thwarted. So in a massive element with hundreds of protons and neutrons, where the nuclear particles are barely hanging on, a chunk of the nucleus can spontaneously break off. If a neutron flies away, and if the surroundings are sufficiently crowded, it may crash into adjacent atoms and knock off a neutron or two there too. In no time (actually, a millionth of a second), you can have a billion trillion neutrons frantically flying and crashing, each emitting energy.

The emissions experienced by victims of radiation accidents amount to free neutrons flying through the body, which isn't good for you, along with flashes of gamma rays, the most energetic form of electromagnetic radiation, which, similarly, no one would seek out as any sort of tonic. Gammas travel at light speed and will break apart atoms and molecules like a cue ball striking a rack of billiard balls. These are cataclysms of their own even if they lie in a realm too small for even microscopes to visualize.

A nuclear reactor is a place where breakups of heavy atoms like uranium-235 and plutonium-239 are encouraged, with energy released in the form of heat. The number after the element's name is simply the total number of particles in the nucleus. For example—and bear with me if you remember all of this from high-school physics or chemistry—uranium is uranium because it has exactly 92 protons. It also usually has 146 neutrons clinging to its nucleus, making a total of 238 nucleons (protons plus neutrons). This nucleon inventory is why it's called uranium-238, and there's probably some of this material in the earth beneath your home. It's rather unstable, so neutrons and even protons can spontaneously break off, though you wouldn't want to sit around waiting for it to happen. Every four and a half billion years, half of the uranium in any sample will lose enough nucleons to decay to a stable form of lead.

If you're in a hurry to see uranium break up, then you'll want to play with a somewhat unusual form of it that makes up only 0.7 percent of Earth's natural uranium inventory. This is uranium-235. It has the mandatory 92 protons so it can be called uranium, but it possesses only 143 neutrons. It is a particularly unstable atom. If hit with a stray neutron, it will cough up two of its own, and each of these will fly at nearly light speed into another atom of nearby uranium-235, which will in turn lose two neutrons, and such a geometric progression ensures that in short order, you've got neutrons flying everywhere.

Again, every element has an established number of protons in its nucleus. Hydrogen has one, oxygen has eight, carbon has six, and uranium has ninety-two. There's usually a matching number of neutrons in the same nucleus, so an ordinary oxygen has eight neutrons and an ordinary carbon has six. But as we saw in chapter 3, some rarer forms of oxygen or carbon have a different number of neutrons. Each variety is called an isotope, and it's labeled, as we've seen, by the total number of nucleons in its nucleus. So, staying with carbon a bit longer, its most common form is carbon-12 because it has six protons and six neutrons. But one of every trillion carbon atoms in the CO_2 in the air you breathe has two extra neutrons, and this is carbon-14. Such rarer isotopes are often unstable and emit radiation.

The half-life of carbon-14 is 5,730 years, which means that in that much time, half of any sample of it will change into something else. Such a long half-life implies that at any given moment, it's emitting little or no radiation.

Or consider uranium-235. With a half-life of seven hundred million years, it's not terribly radioactive either. But radium, whose most stable isotope is radium-226, has a half-life of sixteen hundred years. Since this is much quicker than the other atom varieties just mentioned, we would correctly infer that it's more radioactive and hence more dangerous to handle. In any case, for nuclear power production, uranium-235 is often used, mostly because when it's hit by a stray neutron, it fissions easily, perpetuating the chain reaction.

Whether you get intense heat and radiation capable of being channeled into steam and turbine operations for electrical-power generation or a city-destroying explosion depends on several important things.

First, the concentration of the fissionable material. A power plant's fuel needs to have its U-235 concentration enriched from the normal 0.7 percent found in natural uranium to 4 percent. But this same 4 percent would never explode; for that you need a purity above 90 percent.

Second, the neutrons released by fission are far too fast to create further fissioning in a power plant. They need to be slowed down a millionfold, a process called moderating. In some power plants, graphite rods do the moderating, but in the majority, ordinary water does the job, and the water also serves as a coolant.

The third factor is the uranium-235: the quantity of it, how densely it's packed, and even what shape its mass is in. An amount of U-235 that is barely critical (that is, able to sustain a chain reaction of continued neutron releases) can turn supercritical at an undesirable moment if some carelessness unfolds in its handling. Sadly, this has happened repeatedly.

For example, uranium is sometimes kept mixed with water. Given a safe subcritical storage, like a long, thin cylinder five inches wide and ten feet high, this solution might be safely pumped through pipes in a reactor until it's dumped into a spherical container. Spheres are the most perilous shape in the reactor world. That's because a globe has the smallest surface area of any three-dimensional shape, which means its contents are the most concentrated. In a spherical vessel, every uranium atom is close to the maximum number of other uranium atoms, and a well-behaved subcritical quantity of uranium in a skinny vessel can suddenly go supercritical when placed in a fat cylindrical tank of the correct dimensions, such as a squat soup-can configuration.

This can happen by accident. On December 10, 1968, in Mayak in Soviet Siberia, a worker unthinkingly poured four gallons of uranium water into a large pail. In a microsecond, the uranium atoms, which had been safely diluted simply by being in a tall narrow vessel, were now close enough to one another and in sufficient quantity to start a cascade of neutrons. This criticality of a neutron chain reaction begins so abruptly—with billions of trillions of atomic breakups or neutron releases in far less than a second—that the only sign of something amiss is a telltale blue flash of light as the room's air atoms are broken apart or ionized.

On that cold day, an ordinary pail had become a nuclear reactor. The fierce sudden heat from all those atoms suddenly fissioning violently boiled off much of the uranium water and flung a couple of gallons into the air, onto nearby workers, and onto the floor. (Taken to the hospital, the worker who'd started the reaction survived—barely, having been exposed to seven hundred rem of radiation, an intensity that is often fatal.)[2]

But the peril wasn't over. A bit later that night, the reactor plant supervisor arrived at that basement room to try to determine what had happened. He saw the heavy bucket on a shelf and innocuously moved it to the floor. Thanks to the boiling having splashed away some of the pail's uranium water, the pail's contents were now at subcritical mass and safe. But merely by setting down the bucket in the shallow pool of spilled uranium water on the concrete floor, the supervisor had unwittingly placed the uranium close to more of it. The mass went over the edge and back to a supercritical state. Suddenly there was sufficient uranium mass in a small enough area to start a reaction, and when it again went critical, it was accompanied by another blue flash. In an instant the supervisor absorbed 2,450 rem of radiation, enough to kill anyone twice over; he died two weeks later.

The same sort of accident happened on September 29, 1998, at Japan's JCO nuclear plant. This time the culprit was forty-five liters of an 18 percent enriched uranium-235 solution. It was in a large tank that had, at its bottom, two bladelike propellers that could be stirred by an electric motor. The uranium water was floating atop a solution of a denser fluid, and the worker was supposed to mix them. He draped his body over the large tank so that he could look

183

through a glass peephole viewer, reached over, and hit the button. The blades started turning. Suddenly there was a bang and a blue flash visible throughout the plant. The thirty-five-year-old operator felt unwell immediately and soon started vomiting. He had received seventeen hundred rem of neutrons and gamma rays, and he died eighty-two days later of "multiple organ failure."

That cataclysm of a hundred billion trillion fission events compressed into a single second also zapped a nearby forty-year-old co-worker with lethal high-speed neutrons, though he managed to hang on in a hospital for seven months before succumbing. The tank-turned-nuclear-reactor event also delivered significant but sublethal radiation doses to four hundred and forty other plant workers.

Why did this happen? The answer is fascinating. When the motor started the blades turning and the water stirring, the uranium water's surface formed the shape every coffee drinker observes when stirring her cup, a kind of miniature whirlpool with a depressed center. It so happens that this configuration allowed more uranium atoms to descend into that whirlpool and be closer to the other uranium atoms in the liquid. The change in fluid shape was enough to bring the barely subcritical uranium into a state of prompt criticality. Flash!

The half a dozen similar reactor accidents that produced fatal radiation releases in the United States, Great Britain, Japan, and the Soviet Union between 1950 and the 1980s mostly stayed below the radar of the mass media. This was intentional. The public's stance toward radiation ever since the early years of X-rays and fluoroscopes has varied between wary fascination and terror. Governments that actively invested huge sums in trying to develop and perfect nuclear electrical-power generation just as actively tried to keep the lid on accidents.

It didn't help that nuclear-safety experts were themselves unsure how much radiation was harmful. After all, everyone receives background radiation. Some 360 annual millirem for each of us is normal and probably harmless. We get it from the ground, from bricks and stones, from dust from distant coal plants,[3] and from cosmic rays, with the greatest intensity for those who live at high elevations. We get it from certain foods; a single banana delivers more radiation from its radioactive potassium-40 than a person will receive from living next door to a nuclear power plant for an entire year.

We get lots of it if radon gas leaks through basement cracks from the ground under our homes. We get some every time we fly in a jet—six millirem per round-trip coast-to-coast flight. The question is, How much is too much?

Until recently, health professionals trusted the linear-no-threshold (LNT) model, which says that all radiation is harmful and there is no lower limit to what can hurt you. In other words, though everyone agrees that one thousand rem will kill you, and four hundred and fifty rem has a 50 percent chance of killing you, most health professionals felt that a ten-rem dose might still produce the kind of DNA damage that would be expected to eventually kill one person in ten thousand.

If so, then many nuclear accidents are cataclysms, for if a mere ten rem was inadvertently released to bathe a large area of one hundred thousand residents, a hundred of them might eventually die. Not good.

But there was a problem—this wasn't what seemed to be happening. Studies of large populations who had received low radiation doses showed, after decades, that far fewer of them were coming down with cancer than the LNT model predicted. Epidemiologists monitored Hiroshima and Nagasaki survivors and Chernobyl victims and did a decadelong study of seventy thousand residents near a radioactive, thorium-emitting black-sand beach in Kerala, India, and found cancer rates low. In a few cases, they were lower than in the control groups, leading some to conclude that very low radiation doses are good for you, a concept called radiation hormesis.

In 2005, the prestigious *Journal of Radiology* ran an article that concluded, "The linear-no-threshold (LNT) hypothesis for cancer risk is scientifically unfounded and appears to be invalid in favour of a threshold or hormesis. This is consistent with data both from animal studies and human epidemiological observations on low-dose induced cancer. The LNT hypothesis should be abandoned and be replaced by a hypothesis that is scientifically justified."

More studies are under way. One of them is putting animals in a zero-radiation environment so that they're shielded from even the normal earthly background of 360 microsieverts per year. Will they die sooner when they're deprived of all radiation?

It's too early to know. The biggest health organizations continue to advise people to avoid unnecessary radiation, and hormesis remains controversial.

This matters here because if an accident at a nuclear power plant releases small amounts of radiation into a wide area, as happened in 2011 around Japan's Fukushima nuclear power plant, it might be labeled a cataclysm if many cancer deaths descend like a plague on the surrounding population in the subsequent decades. But if no one is harmed or if the region's cancer rate is actually *reduced* by the event, then the label *cataclysm* vanishes.

In any case, how many deaths make an event qualify as a cataclysm? Every year in the United States, coal-fired power plants produce around twelve thousand deaths from respiratory illnesses such as emphysema and lung cancer. If these same twelve thousand annual deaths had resulted from a single nuclear power plant accident, it would exceed the total eventual death toll from the Chernobyl event, and it would certainly count as a cataclysm. But as things stand, those annual twelve thousand coal-power deaths are pretty much ignored by the media and the public. No headlines, no attention at all.

Since a far smaller casualty toll, like one that might kill just 1 percent of that coal-power number in a hypothetical nuclear plant mishap, would be a headline-grabbing disaster, it's clear that nuclear accidents have their own separate cataclysm standards as far as the public is concerned. It is largely for this reason that we include them in the following pages.

The first of these headline-makers had an unfortunate PR prelude. It was almost as if some malevolent anti-nuke group had set the whole thing up, because the negative impact to nuclear power could not have been more devastating.

CHAPTER 25

A NEW STATE CAPITAL?

On March 16, 1979, a mere twelve days before Pennsylvania's Three Mile Island nuclear accident, the movie *The China Syndrome* was released in theaters to immediate high ticket sales. The film, starring Jane Fonda, Jack Lemmon, and Michael Douglas, was intended to exploit the public's wariness about nuclear power.

The public had—and still has—a very low understanding of the atom and of how power is extracted from it. Many think that an accident at such a power plant could result in a thermonuclear explosion. It didn't help that the nuclear industry, as we've seen, had a long series of sporadic, largely unpublicized accidents where just enough information had leaked to maintain a general air of unsettling mystery and official untrustworthiness.

Moreover, the nuclear industry had spent most of the 1950s trying out various reactor designs. In the early days, meaning the 1950s into the 1960s, companies built full-scale experimental plants that used liquid sodium metal for cooling and breeder reactors that were intended to produce their own plutonium fuel as a by-product of generating power. Melted-salt reactors were tried, as were heavy-water systems and several other designs.

By the 1970s, an early configuration that had been advocated in the 1950s by U.S. Navy admiral Hyman Rickover for nuclear-powered submarines seemed to be the most reliable, and it appeared that it would ultimately prevail over the other systems. This design used ordinary water to cool and control the reactions. (Actually, the ordinary water is called *light water* because its hydrogen atoms have no neutrons. The term differentiates regular water from heavy water—deuterium and tritium—which later fueled the H-bomb.) The coolant was critical because a useful atomic-fission reaction, usually sustained by uranium pellets or rods enriched to about 4 percent of the isotope U-235, creates around 1,100 degrees of heat in the most typical configuration. This heat is used to boil water into steam, which turns turbines to generate power.

All will be well as long as the excess heat is continually channeled away. But if any problem prevents the uranium fuel from being fully submerged in circulating coolant, all hell breaks loose. This was the challenge in the new pressurized-water reactor (PWR) nuclear power plants, which were then built by only a handful of companies, specifically Babcock & Wilcox, Westinghouse, Combustion Engineering, and General Electric. The Three Mile Island facility had two side-by-side plants, both B & W model 177FA reactors. Each used a hundred tons of enriched uranium divided into 36,816 fuel rods. The reactor's pressure vessel had steel walls nine inches thick, and the two seven-story generators created enough steam to produce 1,700 megawatts of electricity.

In March 1979, the first reactor, TMI-1, was shut down for routine refueling, and TMI-2 was operating at 97 percent full power. The plants sat on a three-mile-long sandbar in the Susquehanna River just outside of the state capital of Harrisburg. (The author has flown above it in his four-seater plane several times, with air traffic control's permission. It still looks impressive and is actually impossible to miss, as some of the excess steam-water is continuously released by chimneys atop the iconic thirty-story towers, which are designed to cool a million gallons of water a minute.)[1]

But at four in the morning on March 28, 1979, exactly one week and five days after *The China Syndrome* hit the nation's theaters, a small problem arose.

All steam-water that was not vented out those enormous chimneys was recycled back into the plant, but it first had to be cleaned of any small impurities picked up as tiny pieces of pipe residue, and this was accomplished by passing the water through eight twenty-five-hundred-gallon tanks filled with purifying resin balls. These beads had recently become gunky and stuck together, and plant operators were having no luck trying to clean them with compressed air blasts. The reactor's pressurizer was now leaking a bit too, which was common and not serious. The pressurizer's job was to maintain a mass of air in the closed system, which allowed its pressure to safely vary with changing conditions. It was necessary because the water coolant, being incompressible, could otherwise transfer any pressure shock of a sudden change, like a routine valve closing, to far-flung pipes or pumps, inflicting damage. All plumbers are familiar with the issue, commonly called a hammer, experienced as a pressure or shock wave that makes ordinary water or radiator pipes clank or bang.[2]

The gunky resin beads and the slight pressurizer leak were minor issues, and no one on duty deemed them to be worrisome. But someone on the previous shift had forgotten to close a valve near the pressurizer, so water was now getting into the air lines that were necessary to maintain the correct pressure. Eventually enough fluid accumulated so that a routine closing of a valve created a high-pressure water hammer that clangily traveled through several pipes and destroyed a pump. Little problems were starting to add up.[3]

Alarms and emergency systems were activated in the control room, and a turbine now received a computerized warning that it would soon lose the steam pressure that fed it. In three seconds the reactor automatically slammed home all its neutron-absorbing control rods, shutting down the reactor. At this point, it might seem that all should have been well. But it was not.

In a nuclear reactor, a total shutdown does end electrical generation but doesn't stop the heat from fission reactions because secondary nuclides are created during normal high-power operation. These have various half-lives, lasting from minutes to a few hours, and will keep fissioning for a while until stable fission end products are reached. During this period, the reactor core continues to generate fierce heat and must be kept actively cooled. To use real numbers, the TMI reactor models go from their shutdown heat level of 247 megawatts to 57 megawatts after one hour. Thus, if the core can be kept cooled for just that first critical hour after shutdown, a cataclysm will be averted.

A supervisor asked an operator for a temperature reading and the worker shouted back, "Two hundred and twenty-eight degrees"—Celsius—which indicated normal or near-normal conditions. Unfortunately, the man had misread the gauge or had misspoken, because later records revealed that it was actually showing 283 degrees, a significantly high abnormal value. The supervisor, thinking everything was within normal limits, stopped troubleshooting and missed the fact that a stuck-open valve was causing all of the reactor's primary coolant to boil away.

During the next hour, the critical first hour after shutdown, thirty-two thousand gallons of coolant had been lost. The coolant level had dropped dangerously. The uranium fuel was now totally exposed to air and lacked any cooling mechanism. Its heat kept rising until it reached five thousand degrees, which is twice the temperature of the melting point of steel. No nuclear power plant in history had ever gotten this hot.

They were in new territory, as the uranium was now melting and releasing radiation into its surroundings. The plant's supervisors did not yet know it, but at this point, the entire billion-dollar reactor was lost and would never work again.

The following twenty-four hours were a time of high drama and anxiety. Hydrogen gas was created by the intense heat and by chemical reactions from various melting compounds combining with the remaining steam and boiling water. The hydrogen rose and started concentrating at the containment building's curved ceiling. With all vents closed, its pressure built up to two thousand pounds per square inch. The gas concentration kept accumulating. In a few hours it reached the astonishing pressure of twenty thousand pounds per square inch, which was off the scale on all the gauges and instruments. The building's radioactivity was going off the scale too.

Plant supervisors wondered if the building would rupture and release radioactive gas and dust into Harrisburg, if the hydrogen at the ceiling would explode, or if both would happen. The public, now informed by radio of an emergency at the plant, fled from the region, concerned that a nuclear bomb was about to detonate in their neighborhood. By the end of that weekend, 135,000 people had evacuated the area.

On *Saturday Night Live*, cast members tried to calm the nervous national audience with forced humor. Comedians Bob and Ray held a contest to name a new state capital. Hollywood's *China Syndrome* publicity had, unfortunately, made the country unusually apprehensive about the hazards of a nuclear accident.

In the end, it was a good news/bad news sort of denouement. The bad news was that the TMI-2 plant was a total loss. A third of its uranium fuel had melted, and 70 percent of the core's structure lay in tangled ruins. Cleaning the radioactivity and rebuilding was not even close to being economically worthwhile. The accident, coupled with the recent movie, was the last straw for U.S. nuclear power generation. Since then, nearly forty years ago, not a single new nuclear power plant has been completed in the United States.

The good news was that the hydrogen never blew up. That element requires a prodigious quantity of oxygen in order to explode, and there was no oxygen near the ceiling. Indeed, any free oxygen in the containment building would have been captured by the melting uranium to form uranium oxide. And the containment structure held; it did not rupture or leak. In fact, the only significant radioactive substance released into the air was xenon-133.

Happily, xenon is an inert gas that sits on the periodic table alongside argon and neon, all of them unfriendly "noble" elements that do not combine with others. The body has no use for xenon. Breathe it in, and nothing bad happens; you'll simply exhale it. Health experts started carefully monitoring medical records in the region, especially in areas east of the plant, the direction toward which the wind was blowing that weekend. In the nearly forty years since the accident, no rise in cancer incidence has been seen.

So was TMI-2 really a cataclysm, or did the public merely perceive it as one? This issue arises occasionally in this volume, and it is all the more relevant when it involves events that unfolded during our lifetimes. After all, in that same period, the Ebola epidemic was considered a crisis in the United States, even though it ultimately killed only two people here.[4]

But the TMI accident failed to claim even a single victim, according to nearly all credible academic analyses. Perhaps in only one sense it qualified—it almost single-handedly provoked public opposition to nuclear power, especially when people were roused by a series of antinuclear demonstrations across the country later in 1979.

The largest drew a crowd of two hundred thousand in New York City in September 1979. People heard speeches by Jane Fonda and Ralph Nader and attended a series of highly publicized No Nukes concerts at Madison Square Garden. All this unfolded after a 1978 march and antinuclear rally in Washington, DC, that drew an estimated sixty-five thousand marchers, including California governor Jerry Brown. With this sort of groundswell and public furor, even President Jimmy Carter, a nuclear engineer himself who'd told his staff after visiting the damaged Three Mile Island reactor complex that the accident had been "minor," was intimidated. A few years earlier, the president would no doubt have reassured the country that nothing much had happened and that nuclear power was perfectly safe. But Carter couldn't ignore the handwriting on the wall, and he did nothing to try to counteract the tide of popular opinion.

Thus Three Mile Island, though it remains an official disaster in some published historical chronologies, was indeed the single accident that effectively killed nuclear power and thereby stifled a major method of electricity generation, the only one besides hydro that can produce steady, round-the-clock, carbon-free electrical power.[5]

If climate change ends up following one of the worst-case scenarios feared by climatologists, then the TMI-2 "cataclysm" may indeed end up deserving that moniker.

CHAPTER 26

SECRETS OF CHERNOBYL

To appreciate the Chernobyl cataclysm, one must first understand its RBMK reactor—a crude design that was original to the Soviet Union and not merely a copy of Western ideas. As everyone knows, RBMK stands for Реактор Большой Мощности Канальный, which is pronounced "Reaktor Bolshoi Moshchnosti Kanalnyy" and in English means "high-power channel-type reactor." (Yes, the Chernobyl reactor's name contains the word *bolshoi*, which might conjure images of leaping ballerinas. Forget that. The word in Russian means "great" or "grand" or "big," and the label was appropriate for such a grandiose, high-power device.)

The RBMK had a number of inherent design flaws that were largely responsible for the 1986 disaster, although the precipitating cause was a very poorly conceived test by supervisors who lacked the knowledge to conduct such an operation and who then repeatedly mishandled the situation as things went awry. Nonetheless, several of the RBMK's greatest deficiencies were fixed after Chernobyl, and eleven reactors of the same design continue to produce electricity in Russia to this day.

The RBMK was intended to be cheap and efficient in a number of different ways and to produce plutonium as a by-product, which would then be used for the Soviets' nuclear-bomb production. It used ordinary filtered river water for cooling and standard inexpensive graphite rods for controlling the fission reactions. Natural unenriched uranium fueled the plant rather than the precious uranium isotope U-235 favored in Western designs, as U-235 was expensive to produce. All these features, as well as the absence of a concrete-and-steel containment building in case of a major accident, yielded the type of enormous reactor that could be built cheaply and in large numbers. In Lithuania, an RBMK reactor, which even today is one of the world's largest, continues to produce 1.5 gigawatts of power.

RBMK Reactor Design

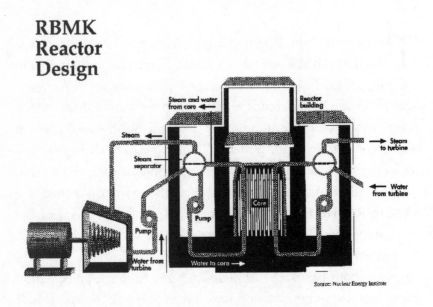

Source: Nuclear Energy Institute

(Argonne National Labs)

This design offers several undeniable benefits. For example, as we've seen, the first hour after a reactor shutdown is a perilous period when coolant must still be circulating in the core. However, the RBMK has an unusually low amount of fissions per cubic meter of its core, which lets it withstand a total loss of electrical power, including a complete circulating coolant failure, for up to an hour with no expected core damage. (If the later Fukushima reactor had had such a strength, it might not have been destroyed in the 2011 accident.) Unfortunately, more than balancing out that benefit is the fact that in an RBMK reactor using uranium as fuel, the circulating water doesn't merely serve as a coolant; it also moderates the fission reactions and dampens them to safe levels. This is exactly the opposite of what water does in a Western B & W pressurized-water reactor, where the neutron production critically relies on nearby water molecules to proceed.

So when you lose the circulating cooling water in a Western power plant, bad things happen, but at least the fissioning power production automatically shuts down. Conversely, in an RBMK plant, the lack of water makes the fission process *increase*—and this runaway is exponential. It's a double whammy and quickly makes the power and the heat rise radically. Known technically as having a positive void coefficient, this property of an RBMK is no small deficiency. It was central to the cataclysmic events of April 26, 1986.

But other design-safety aspects were wanting as well. The only thing blocking radiation from escaping the core was a massive, round, eight-foot-thick concrete cap on the top of the five-story-tall reactor. High above that, there was no outer containment building, merely a tin roof to keep out the rain.

And that still wasn't the end of the problems. Each of Chernobyl's four operating RBMK reactors' power was controlled by 211 enormous, forty-foot-long graphite rods. When a rod was raised or lowered, its middle sixteen feet would make the fission reactions increase or decrease, thus enabling operators to regulate the heat and power. But when a rod was fully lowered in an emergency, the first few feet of it would radically increase the power, so if all of them were thrown down simultaneously, there'd be an initial catastrophic runaway power surge. In other words, operators would have to be calmly judicial when ordering the reactor to shut down, and they would have to extend rods a few at a time manually, as if it were all a video game. This was very different from Western reactors, where a single red Scram button punched by a frantic supervisor will drop all control rods within three seconds, abruptly stopping the power production, end of story.

With all these system quirks and design negatives, and with no fewer than four RBMK operating reactors and two more under construction just eleven miles outside the ancient historic Ukrainian town of Chernobyl, and with the new supporting village of Pripyat with fifty thousand people a mere 1.9 miles from the plant's gates, you'd think extreme caution would be the daily buzzwords in the reactors' operations. Yet in April of 1986, although all the supervisors and foremen were mechanical engineers and experts in areas such as turbines and wiring, there was not one person on the site who was deeply knowledgeable about graphite-moderated nuclear reactors or fully versed in nuclear theory.

And that's when the men in charge decided to shut down all the plant's safety systems to see what would happen.

Wait, scratch that; although it's perfectly true, such a summary makes them seem unduly stupid. There was actually some logic behind that night's protection-disabling idea. Essentially, their safety committee had long ago realized that in any kind of emergency shutdown, there would be some lag time between when their generators stopped and when they could get the diesel backup generators going. In that interval, the coolant pumps wouldn't be running, and the buildup of heat in the reactor could cause damage, even though their system should theoretically not succumb to a full meltdown. But some engineer had theorized that at least one of the massive turbine rotors should keep spinning long enough from its own momentum to just barely be able to power the coolant pumps during those critical few minutes. A previous test had yielded pessimistic data, so they'd figured out a way to reduce the drag caused by the generators' electrical fields and now wanted to run the test again. Thus there was some genuine purpose to this experiment.

In preparation for it, an engineer named Gennady Metlenko installed a special control-panel switch and had it labeled MPA, which stood for Russian words that meant "maximum design basis accident." Essentially, anyone hitting that switch would instantly unleash the worst thing that could possibly happen. All at once, the turbine would be shut down, as would the emergency cooling system, including all of the pumps. The switch also blocked the diesel generators from starting. It would bypass all the automatic safety controls and disable everything that kept the reactor running properly.

In his analysis of the accident thirty years later, senior Georgia Tech research scientist James Mahaffey summarized that MPA switch in six words: "This was a monumentally bad idea."[1]

The team of supervisors and foremen began the experiment at 1:00 p.m. on April 25, 1986, when they reduced reactor number 4 from maximum power. An hour later, they disabled the emergency cooling system to prevent a hundred thousand gallons of cold water from crashing into the red-hot reactor and perhaps warping something.

Then an unexpected power demand from far-off Kiev made them postpone the test, and they temporarily powered it back up to maximum. They resumed their experiment at 11:10 p.m., but instead of the power reducing and holding at fifteen hundred megawatts, as planned, the reactor kept running down all the way to a mere thirty megawatts. They then correctly realized that the unintended spin-down was due to a fission-power sequence of radioactive isotope decays, which had disturbed the reactor's normal equilibrium.

The engineer in charge of reactor number 4, Anatoly Stepanovich Dyatlov, well known for his temper and for his minimal understanding of operational nuclear fission, realized that the reactor was now stuck at a low power level and might require two full days to be restored. As the person leading the test, he told his employees to bring the power back up fast, then he paced, yelled, cursed, spat, waved his arms, and threatened the workers who were protesting that it was against protocol to bring the power back up too quickly. An operator, Leonid Toptunov, fearing for his job, began hitting switches that pulled out control rods until he'd managed to yank out 205 of the 211, which brought the power up to two hundred megawatts. But then, at 1:22:30 a.m., the operations computer sounded an alert warning that in the reactor's current configuration, control might be lost; it announced that the reactor should be shut down right away.

Dyatlov wouldn't listen. In another two or three minutes, everything would be fine, he shouted. Thirty-four seconds later, he and his team began their experiment. One turbine was shut down, and an operator hit the perilous, bizarre, newly installed MPA button.

The nearly red-hot reactor immediately responded by violently boiling away its coolant water. With the water disappearing, the fissioning increased explosively. The power level zoomed upward. As the cooling water disappeared, the reactor suddenly attained a state of supercriticality, where fissioning increases exponentially. In the control room, the power-level needle moved to its highest level on the gauge while an operator stared at it in disbelief for thirty-six seconds.

Then the stunned operator shouted a warning and, not even asking his boss for permission, hit the red button that dropped all the control rods, trying to kill the runaway reactions. The reactor went "prompt supercritical," where neutrons all fly at once and don't have to wait for any response from neighboring atoms. Now the reactor core started to melt. Vertical pipes that normally guided the graphite rods warped and twisted, rendering further control impossible. The reactor's power level geysered upward to an unbelievable thirty *billion* watts as the structure began disintegrating.

The jammed graphite rods in the middle of the core did more than merely prevent further human control or amelioration; they created a positive loop, which actually boosted the activity of the uranium fuel. This was later deemed the "final trigger" of the day's catastrophic events.

The building shook, and the unprecedented high temperature boiled away every remaining ounce of coolant water, producing steam under enormous pressure. When the steam exploded a few seconds later, the five-hundred-ton concrete cap on the reactor roof blew into the air. Fragments from the reactor tore upward like shrapnel, destroying the tin roof.

All the plumbing lines into the reactor now came loose and broke, and water, instantly changed to steam by the intense heat, mixed with radioactive dust and debris and blew past the torn roof and out into the sky—the first moments of a giant fallout cloud that would soon envelop all of Europe.

A buildup of hydrogen gas, now refreshed with a sudden influx of oxygen from the missing roof, detonated in an enormous explosion of its own. The entire Chernobyl number 4 plant, already nearly totaled by the events of the past few minutes, was completely totaled this time. Vaporized metals, zirconium, graphite, and fifty tons of uranium fragments were hurled thirty-six thousand feet into the sky where they contaminated every commercial airliner within a hundred miles. The plant's remaining graphite rods, eight hundred tons in all, started burning with intense flames.

The experiment was apparently over.

In the control room, the violent explosions had turned the concrete walls into rubble; sparking ceiling fluorescent fixtures now dangled from fallen electrical wires. All the gauges and instruments were dark and dead. Battery-powered emergency lights cast small spots of brightness here and there. Chernobyl number 4 had ceased to exist.

Radiation was intense, but the handheld monitoring instruments, designed to detect only reasonably high leaks or emissions, all had needles stuck uselessly flat against their upper limits. Even specialized monitoring equipment would have been unable to read the unbelievable levels; thirty thousand roentgens per hour was zapping out of the burst-open reactor. Pieces of previously flying fuel and graphite were littering the plant, and each of these chunks was emitting five thousand rem, five times a fatal dose to anyone who paused in their vicinity.

Meanwhile, Dyatlov, arrogant and ignorant as always, believed the inside of the reactor was unharmed and that all would be fine if cold water could be pumped in. He even said so in a call to Moscow, in which he assured his superiors there was nothing to worry about. Firemen arrived and rushed to the roof where the tar was burning furiously. No one warned them about the radiation hazard, and few of them lived to the end of that spring.

More than one hundred graphite rods remained furiously on fire. Blazing updrafts from this inferno continued unabated for the next nine days, and these lifted plumes of intensely radioactive debris from the torn-open reactor straight up into the atmosphere. The thin metal roof was now a grotesque scene of enormous gaping holes with more empty space than metal directly above the reactor.

In all, a hundred and twenty-seven control-room personnel and firemen came down with acute radiation sickness, and thirty-one died. Later, twenty-three more people in the immediate area perished from radiation effects. The radioactive cloud spread northward, first to Finland and Sweden and then to the rest of Europe, but the Soviet government warned no one, having considered Chernobyl a secret facility. The contamination from iodine-131, strontium-90, and cesium-137 was predicted to produce four thousand to ten thousand cancer deaths, and as of 2018, according to the Ukrainian government, around eight thousand have already perished.

Of the dozen people in the control room, five died agonizing deaths from radiation burns in the days immediately after the disaster. Anatoly Stepanovich Dyatlov received a dangerously high dose of radiation and was left a permanent invalid; six years later, he couldn't walk more than a few steps without exhausting himself.

According to the official Soviet report, Dyatlov's incompetence was directly responsible for the disaster, and soon after the catastrophe he was found guilty of criminal negligence and sentenced to ten years in prison, though he and the other convicted Chernobyl supervisors were pardoned after serving six years. The official version of events states that Dyatlov violated the most elementary safety precautions on the night of April 26, 1986, and bullied his subordinates into taking unnecessary risks that led directly to the destruction of the reactor and the spewing of radioactive particles across a wide area of Europe.

In terms of the amount of contamination it produced, the Chernobyl explosion was equivalent to more than ten of the atomic bombs dropped on Hiroshima, and hundreds of thousands of people were permanently forced from their homes. It was by far history's greatest radiation accident.

In rankings of all industrial mishaps and in terms of sheer number of deaths, Chernobyl may well have been history's very worst in this category as well. Factor in the fright-producing component of it happening in a nuclear power plant, and we've got the top candidate for an unnatural event that merits that big word on the cover of this book.

CHAPTER 27

THE HYBRID CATACLYSM

The cataclysms we're exploring are mostly of natural origin. Our human hands are clean when it comes to supernovas and tsunamis. And although the jury may still be out on the success of the Big Bang—which gave us parakeets but may ultimately crush the cosmos to the size of a dog biscuit—we members of *Homo bewilderus* can shrug it all off with a "Don't blame me, I wasn't even there" innocence.

We are also exploring several full-scale disasters of our species' deliberate design, among them World War II and the hydrogen bomb. In the past few chapters we've looked at the lovable goofs and follies that led to nuclear power plant cataclysms. But now, for the first and only time, we visit a mongrel cataclysm of mixed pedigree.

The 2011 Fukushima event was part natural and part human mea culpa.

The natural component took by far the most lives. In fact, the man-made portion has yet to kill a single person. Yet it is the latter that the public remembers and that produced the most dramatic consequences. Assembled as a cohesive narrative, the story seems more astonishing now than it was that Friday a few years ago.

On March 11, 2011, the strongest earthquake ever to strike Japan shook the country in the early afternoon. Measuring a

Richter magnitude 9.1, the Tohoku quake was among the top five strongest earthquakes recorded on the planet. It was caused by a violent undersea tectonic-plate shift in a subduction zone just forty-three miles east of Japan, and it abruptly moved that country eight feet closer to Hawaii while sinking it more than two feet. This in turn sufficiently changed the planet's mass distribution to speed up its rotation, reducing the day's length by a few microseconds. This was one of the event's few positive developments, as Japan's residents were no doubt happy to see that day end as soon as possible.

An enormous tsunami resulted from the massive water displacement and reached Japan's east coast just over half an hour later. This tsunami produced all of the casualties that day.

Even now, the event's harrowing YouTube videos are riveting, especially when you see eighteen-foot-high seawalls engulfed by the fast-rising tidal wave. In most places, the ocean rose an astonishing 45 feet. In a few places, the "wave" was 133 feet high, as tall as a fourteen-story building. I put the word *wave* in quotation marks because, contrary to popular belief, a tsunami is not a single large wave that wreaks destruction. Indeed, in most places, the first disturbances in the water that day were small ripples or minor waves a few feet high. But this proved to be only the leading edge of seawater that steadily rose, inexorably picked up speed, and then easily carried off cars and entire homes.

The tsunami warning issued minutes after the quake was heeded by many but ignored by some. In a few places, people gathered to watch the strangely behaving water from what they thought were safe perches atop seawalls. In mere minutes, however, these proved inadequate, and some eighteen thousand people lost their lives. In many cases, the retreating water swept bodies as well as still struggling people far out to sea.

Such catastrophic loss of life might indeed qualify as a cataclysm. Nonetheless, the day's violence is most popularly associated with the core meltdown at the Fukushima 1 nuclear power plant owned by TEPCO, the Japanese electrical-power company. This was odd by itself, because the billion-dollar damage should not have happened. After all, seventeen other nuclear power plants were operating at full power the moment the earthquake struck, and several of these were even in the same region as Fukushima 1. Yet all of them were safely powered off to full cold shutdowns with no damage and no radiation leakage whatsoever. What, then, happened at Fukushima?

The result of a strange sequence of mishaps at that single facility, still largely unknown to the public, created the only consequences now generally remembered. If the whole rationale for global nuclear power vanished in a puff of white smoke that day, along with the reputation for safety of American-designed nuclear plants, and if that did indeed spur Germany and Japan to shut down all their own safely running nuclear plants within a few short years despite their governments' knowing that such non-carbon-producing electrical generation was vital in the fight against climate change, then the sum of all the consequences was indeed cataclysmic. We obviously need to know what really unfolded that afternoon. All of Japan's other power plants weathered the earthquake and tsunami; what happened at Fukushima 1?

It wasn't the earthquake that caused the problem. Historic though it was, and rough as it was on the myriad components of Japan's complex reactors, they all survived those 2:34 p.m. jolts that were violent enough to make standing impossible. At that point, the automated systems at all of Japan's nuclear plants initiated shutdowns. As we've seen, the fierce heat of a billion-megawatt reactor operating at full power does not abruptly end with a flick of a few switches. When reactors are running, highly radioactive intermediate nuclear materials are continually created, and these keep fissioning for hours and even days as their half-lives convert them to stable substances. As we saw in the previous chapter, the first hour after shutdown is a particularly critical time to maintain coolant to the blisteringly hot core as the fissioning afterglow slowly diminishes.

After an emergency shutdown, coolant pumps are run by external AC current from incoming power lines, but on March 11, the quake knocked out electricity throughout the country, and all the plants' external power feeds were dead. No problem; power was then supplied by backup diesel generators, and at Fukushima, all twelve of these started roaring away before even one minute had passed. All was well. And the isolation condensers, designed to provide cooling water after a shutdown, were operating normally at each of Fukushima's six reactors.

But an operator at the Fukushima reactor number 1 thought his reactor was cooling too rapidly. Bewilderingly, and apparently overthinking the situation, he refused to let the computerized system do its job. So this unnamed operator—who deserves to be someday memorialized as a comic-book villain—pulled the handle on a switch that turned off the isolation-condenser coolant. This lever's commands also then shut down two electrically controlled flow valves, MO3-A and MO3-B. As James Mahaffey summed it up in his analysis, "With that simple action, overriding the judgment of the automatic safety system, an operator doomed Fukushima 1 to be the only power plant in Japan that suffered irreparable damage due to the Tohoku earthquake of 2011."[1]

True, plant supervisors and engineers soon realized their urgent need for the coolant. The problem was that most of the valves were electrically operated. They would work as long as the diesel generators were producing backup power. But everything changed when the tsunami hit.

The first wave, arriving forty-one minutes after the earthquake, was thirteen feet tall and was fully contained by the eighteen-foot wall at the beach. But eight minutes later a second wave hit, and then a third. Each was forty-nine feet high, a towering five stories. In a few minutes, the entire nuclear plant was flooded.

Here the first design flaw became spectacularly obvious. At Fukushima 1, all the backup generators in reactors 1 through 5 were in the basement, and so were the emergency batteries. Now, suddenly, these backup generators were underwater. They stopped cold, and thanks to the batteries also being submerged, all coolant pumps halted at that moment as well. The heat from the nearly red-hot reactor core was no longer being carried away. Worse, at unit 1, the condenser-coolant lines, as we know, had been manually switched off, and now there was no way to reopen the electric valves. It was just a matter of time before the reactor's sixty-nine tons of sizzling uranium would start melting.

On the far end of the property, at reactor 6 alone, a single air-cooled generator had been sited aboveground, and this sole source of electricity let units 5 and 6 safely continue their shutdowns.

With even the battery room underwater, unit 1 had absolutely nothing. Even its gauges were dark. The windowless control room was pitch-black, and operators using flashlights could only stare at the dead instruments that told them nothing about the status of the reactor. Tense minutes elapsed. Inquiries were made: Could portable generators be helicoptered in? No, the company couldn't help them.

Three hours after the quake, no water remained to cool unit 1's core; it had all boiled off. Ninety minutes later, the zirconium structures holding the uranium fuel started melting, as did the uranium itself. The four hundred heavy fuel assemblies started falling to the bottom of the reactor, producing steam and newly created hydrogen gas that rose to the ceiling and grew ever more pressurized.

By the next day, March 12, workers knew that an explosion was imminent and that the only hope was letting some of the gaseous radioactive mixture out through the roof vents. They got permission for a release, evacuation warnings were sounded to the neighborhood, and they were set to release the noxious pressurized mixture. But without power, there was no way to open the vents. Heroically, workers rushed to the ceiling areas to hook up jury-rigged hoses and tried to use gasoline-powered compressed air to blast open the valves.

Almost exactly twenty-four hours after the earthquake, the roof valve was at last opened, venting off pressure. Meanwhile, fire trucks had assembled and were standing by with hoses, preparing to pump seawater into the sizzling core of unit 1 to cool it. The salt would destroy the reactor, but it was already destroyed, and this at least would cool the unameliorated residual nuclear fissioning.

It was then that the hydrogen in reactor 1 exploded, hurling radioactive chunks of debris so high into the air, it took quite a while for all of them to land. Though five workers were injured, no one was killed. But the fire hoses were torn and several trucks, cables, and other badly needed equipment were damaged. And radiation now spewed from the wrecked containment building. It thankfully didn't resemble the instantly lethal levels that had been disgorged from the burst Chernobyl reactor a quarter of a century earlier, but it meant that everyone would now require bulky radiation suits.

The next day, the core of reactor 3 melted down too, and at 11:01 a.m. its copious hydrogen gas exploded. This widely televised fireball injured eleven workers and damaged or destroyed backup generators that had been assembled outside the building. The radiation at the opening to the wreckage, some thirty rem per hour, or less than one-hundredth the levels at Chernobyl, was nonetheless the highest measured at the entire plant complex during the accident sequence, and it meant that no worker could remain for more than twenty minutes.

And still the fireworks weren't over. One day later, on the ides of March, everyone got a sudden surprise when the unit 4 containment building exploded. This made no sense, since this reactor had been shut down without incident and all its fuel had been removed and safely stored in an open-rooftop floor tank. It took nearly half a year for analysts to figure out that unit 4 was not the problem. Hydrogen gas had mixed with the radioactive steam that had been vented from unit 3 and gathered in a stack the two reactors shared. When this gas from unit 3 was vented out a duct to the ceiling of unit 4, it ignited.

But nobody knew this at the time. Instead, in the frenzied immediate aftermath of the unit 4 explosion, supervisors started worrying about units 5 and 6, which actually were fine. Remember, these were the reactors that had a backup diesel generator aboveground, and thus the generator had provided power throughout. And nobody at unit 5 or 6 had criminally shut off an emergency condenser cooling system the way someone had done at unit 1.

Nonetheless, supervisors imagined that if unit 4 could blow up for no reason, so might 5 and 6, so workers bravely climbed to the units' roofs and cut crude holes to allow hydrogen gas to escape. There never was any hydrogen gas there, and this worker exuberance was the only damage ever inflicted on those units.

When it was all over, units 5 and 6 would be brought back online without a problem. Unit 4 would be fixable, with relatively small repairs. Unit 1, however, was destroyed, and units 2 and 3 were radioactive and not worth the expense of replacing. A shame; they would have survived if only someone could have replenished the water in their condenser tanks and supplied power. The real villain, other than nature itself, was TEPCO's failure to address the possible hazards of earthquakes and tsunamis. Ignoring warnings they themselves received from seismologists, administrators counted on an eighteen-foot-high seawall to provide 100 percent protection from a tsunami. Then, being certain that the plant would never experience flooding, they sited nearly all their generators as well as their backup batteries in basement rooms.

Fortunately, the radiation releases were all sublethal. One worker at Fukushima received 59 rem total; another got 64 rem. Typical doses to residents living near the plant were 1 to 1.5 rem, which is less than you get during a CT scan. The number of expected eventual cancer deaths from the Fukushima cataclysm varies widely but is always several orders of magnitude below what is expected from Chernobyl. The United Nations estimates that not one death will ensue, and this will be especially relevant if the theory of radiation hormesis is supported by ongoing studies.[2]

On the high end of possible consequences, in 2011, epidemiologist Peter Caracappa at Rensselaer Polytechnic Institute predicted a few hundred extra cancer cases would ultimately appear in Japan.[3] The World Health Organization, using the very conservative LNT model, predicts an increase in thyroid cancer for female infants living near the Fukushima plant at the time of the accident; specifically, a 0.5 percent lifetime increased chance of thyroid cancer. Meanwhile, a June 2012 Stanford University study by John Ten Hoeve and Mark Z. Jacobson deemed that the radioactivity released in the Fukushima sequence could cause 130 future deaths from cancer.

However, annex A of the UNSCEAR (United Nations Scientific Committee for the Effects of Atomic Radiation) 2013 report to the UN General Assembly says that there has been no discernible increased incidence of radiation-related health effects in people living in Fukushima, and none are expected among exposed members of the public or their descendants.

No cataclysm. Except, once again, for the fact that the events of March 11, 2011, permanently sabotaged one of only two methods we have to generate around-the-clock carbon-free electrical power. Indeed, a Japanese poll conducted a few months after the TEPCO accident found that, of 1,980 respondents, 74 percent said that Japan should gradually decommission all fifty-four of its reactors and become completely nuclear-free. In short, the Fukushima accident does indeed appear to be ending that carbon-free method of generating electricity in that nation. A big bright spot in all this is that since then, Japan has committed to a transition to renewable energy sources of carbon-free power production. Half a year after the accident, Japan's richest person, Masayoshi Son, donated one billion yen (thirteen million dollars) to create the Japan Renewable Energy Foundation. Later that same month, Japan unveiled plans to build a pilot floating wind farm with six two-megawatt turbines off the Fukushima coast — a wind farm that became operational in 2015. More are planned.

By 2015, all forty-eight of Japan's nuclear reactors had been taken offline, although the government set a goal of resurrecting a few of them and eventually having a small percentage of the country's power from nuclear. Before the Fukushima accident, 30 percent of Japan's power had come from nuclear; after it, that electricity was produced by fossil fuels, which cost the equivalent of eighty billion dollars between 2011 and 2015. But by May 2016, renewables, including hydro, accounted for more than 20 percent of the country's energy supply.

As with many of our other cataclysms, a phoenix was emerging from the ashes.

CHAPTER 28

THAT THERMONUCLEAR BUSINESS

The biggest bangs we humans have ever created are, of course, Little League explosions on the cosmic scale. Moreover, thermonuclear blasts are not cataclysms in and of themselves. But when we realize that a single one of these weapons exceeds the explosive force of all the armaments of World War II combined, we might reasonably suspect that trouble could be ahead. After all, no historic weapons buildup by antagonistic nations has ever ended with the armaments unused and dismantled.

For a while, it looked like humanity might buck that trend. After the horrific cataclysm of World War II, the United States and the Soviet Union gradually assembled a combined inventory of nearly a hundred thousand nuclear weapons, and, just as worrisome, other nations that had bones to pick began building their own. The planet hovered in an uneasy tension. Would one of the trouble spots, such as nuclear-armed Pakistan and India, erupt following some hotheaded leader's instigation?

We certainly approached a cataclysm during the 1961 Cuban missile crisis. But then came a period of arms-reduction treaties such as SALT and SALT II, and the two superpowers' warheads were reduced to "just" a few thousand apiece. Perhaps, people thought, the world would get through this after all.

And maybe it will. But since there remain minor players with deliverable atomic weapons who seem to lack self-control—in places like North Korea—no one can predict the future. What we can do is see just what these mechanisms are capable of, since the actual lethality of today's surprisingly dirty, fallout-producing thermonuclear weapons would produce truly cataclysmic consequences.

Again, huge explosions do not constitute disasters by themselves. But the current stockpile of peculiar weapons places the planet on a continuous thirty-minute fuse; routine daily life might click over to cataclysm in seconds, and future historians would never be able to figure out exactly what happened.

History's greatest human-designed violent occurrences were the two strongest hydrogen-bomb detonations: the Soviet Union's Tsar Bomba, which was detonated in Novaya Zemlya in 1968, and the United States' Castle Bravo, which was exploded in 1954 in the Marshall Islands. Both had quirky histories.

A thorough review of nuclear-weapons development and their present stockpiles and capabilities lies beyond the scope of this book. But some of these big-boom basics are so amazing they really should be known by all, even those to whom the term *W88* (a common U.S. thermonuclear weapon) means nothing.

We're exploring this for one reason: If a true cataclysm befalls our world during our lifetimes or those of our children, it most likely will not be caused by a rogue asteroid or even a high-mortality pandemic; it will be due to nuclear war or nuclear terrorism. The public was very aware of this possibility in the 1950s and 1960s; nuclear-Armageddon-themed films like *On the Beach* highlighted this global anxiety. That fear also resulted in the birth of several watchdog groups that are still active.

They had reason to worry. In 1960, the U.S. nuclear-weapons stockpile had a total yield of twenty thousand megatons. Somehow, the end of the Cold War and the signing of several treaties that scaled back these weapons—the U.S. stockpile yield of nuclear weapons fell below six thousand megatons by the mid-1970s—led to a widespread complacency that translates today into a vague, general "It won't happen" mind-set.

And maybe it won't. Yet the fact remains that as of 2018, some 14,900 thermonuclear weapons are cocked and primed, deliverable in a matter of minutes. The command to fire can be issued by a single mentally unstable person.[1] Or a computer accident. Or, perhaps, rogue hackers, bored teenagers with nothing more exciting to do than obliterate their planet. Or possibly a deliberate miscalculation by military planners, who might decide that a first strike was in their nation's best interest.[2]

Of course, we all hope it never happens, but in any exploration of cataclysms that includes future possibilities, knowledge is the first step.

Until the 1940s, all explosions required chemical reactions. These always involved position changes in the electrons that shimmer in orbitals above each atom's dense nucleus. But the nucleus itself was thought to be untouchable and stable. It took physicists like Niels Bohr, Emest Rutherford, and Enrico Fermi to demonstrate that atomic nuclei could undergo changes that liberated incredible power.

In 1920, Rutherford showed that the sun shone by such a nucleus-alteration process, of hydrogen nuclei fusing together. The next big breakthrough appeared in a 1939 paper by Austrian physicist Lise Meitner and German chemist Otto Hahn, who showed that the nucleus of uranium, when hit by a neutron, would release an enormous amount of energy as it split into two smaller nuclei, with one piece changed into the element barium. Meitner was the first to use the word *fission* to describe this process.

Enrico Fermi, after emigrating from Italy and teaching at the University of Chicago, created the first *atomic pile,* his term for a self-sustaining, energy-releasing chain reaction. He found that a rare type of uranium—U-235—would release two or three of its neutrons whenever it was hit by a single neutron. These released neutrons would go on to strike other atomic nuclei if there were any nearby, and they in turn would emit more neutrons—and soon there'd be a geometric progression, a runaway reaction.

World War II gave the issue urgency, and Einstein famously signed and delivered physicist Leo Szilard's letter to President Roosevelt warning of the possible development of an atomic bomb by the Nazi regime and the need to prevent the Nazis from being the first to acquire such a weapon. Thus was launched the Manhattan Project, overseen by General Leslie Groves.

The scientists involved eventually discovered that creating a bomb required a fissionable substance that was both purified and collected into a big enough lump. The two practical materials that would best fissile were the synthetically produced trans-uranium element plutonium and the hard-to-extract uranium isotope U-235. They found that a softball-size piece of either would provide enough mass to go supercritical and detonate with the force of over ten thousand tons of TNT.

But in a bomb, such a fifteen-pound quantity would have to be brought together suddenly, by rapidly combining two smaller sub-critical pieces of the element. And its final shape was all-important. If uranium was formed into a cylinder four inches wide, it would remain safely subcritical no matter how lengthy the bar and no matter how much U-235 was present, so no explosion would ever occur. That was because the atoms of U-235 would be far enough apart to prevent a self-sustaining chain reaction. But the same mass of that uranium isotope, or even a bit less, would become immediately supercritical if pounded or pressed into a more compact shape, like a sphere.

Both the plutonium and the U-235 were being culled atom by atom in a multiyear process using high-speed centrifuges in Oak Ridge, Tennessee, and a hastily built plutonium production plant in Hanford, Washington. The nuclear scientists realized in the spring of 1944 that their planned cannon design would not work for a plutonium weapon. New calculations showed that the pluto-nium would start fissioning prematurely, when the first subcritical piece, the "bullet," was still en route to the target chunk of the metal. There'd be no explosion. Or at least not much of one; the fuel would instead get superheated and blow itself apart. Now it seemed that everything depended on uranium-235, and they realized there wouldn't be enough of this fuel for more than one bomb, even by the following summer of 1945.

After years of studies and experiments that included a frantic search to find a new approach for a plutonium-fueled bomb, two workable designs finally emerged. The outcome: At Hiroshima, the 9,700-pound bomb achieved an explosive criticality with a gun-type device in which an 85-pound hollow-cylinder U-235 projectile was fired through a cannon barrel to quickly surround an inner 85-pound solid cylindrical uranium spike that was also subcritical. The resulting U-235 mass, consisting of both the projectile and the target, became instantly supercritical, although only 1.3 percent of the uranium actually fissioned before the material was blown apart. Still, this was enough to create a blast equal to fifteen thousand tons of TNT.

A few days later, in the Nagasaki bombing, the newer configuration was used; it matched the gadget device detonated in the famous Trinity A-bomb test of July 16 at Alamogordo, New Mexico, nearly a month earlier. Here, the 10,800-pound bomb employed an implosion method to achieve criticality of its plutonium-239. The 13.6-pound plutonium sphere, the size of a large softball or small grapefruit, was surrounded by several thousand pounds of high explosives and lenses in the thick steel bomb casing that concentrated the blast inward with great evenness. The plutonium softball was crushed by the implosion to the size of a tennis ball, which instantly caused it to achieve a supercritical state. Although just two pounds of its plutonium fissioned, it created an explosion equal to the force of twenty-one thousand tons of TNT; its initial 12,000-degree fireball matched the sun's surface temperature and was as wide as five city blocks.[3]

Though these atomic bombs were immensely destructive, several U.S. physicists, led by Hans Bethe, Stanislaw Ulam, and Edward Teller, saw these weapons as just a first step, and they started working on a "super," as it was called before the term *hydrogen bomb* became the popular label.[4]

The idea was to use an atomic bomb to create the temperature and the compression of heavy hydrogen (deuterium, which is hydrogen with both a proton and a neutron in its nucleus, and tritium, which is hydrogen with a proton and two neutrons) to achieve a fusion reaction that would theoretically be much more powerful than the original atomic bomb.

It wasn't technically easy. Several concepts were proposed and referred to by various names (Andrei Sakharov named one *sloika,* or "layer cake"), designs that would have layers of fissile uranium or plutonium alternating with heavy hydrogen. But the true 1951 breakthrough, and the secret the United States guarded for decades, was the Teller-Ulam design, which channeled a massive wall of X-rays released by the atomic-bomb explosion to compress the hydrogen and begin the fusion reaction. An alternate explanation of how the Teller-Ulam design works is that the X-rays ablate the

Hungarian-born Edward Teller, often called the "father of the H-bomb." *(Wikipedia)*

bomb casing to create an inward implosive force, a force that was calculated to be 5.3 billion times Earth's atmospheric pressure in the first test device. (The initial pressure inside the bomb casing has been improved to 64 billion atmospheres in current thermonuclear weapons like the W80.)

The big challenge was how to efficiently transfer the energy of the atomic-bomb fission explosion to the hydrogen fuel. Bomb designers introduced what was essentially a third explosive stage; it made the energy-transfer process continue so that each stage ignited and sustained the next. Although the exact method is still a secret, it's widely believed that this interstage energy transfer was achieved using an extremely frothy aerogel code-named Fogbank. All stages in the bomb casing are embedded and suspended in this material.

The first test of the Teller-Ulam concept was conducted on November 1, 1952, on Elugelab Island in the Enewetak atoll of the Marshall Islands in the Pacific. The hydrogen fuel was supercooled liquid deuterium, which had to be maintained by an elaborate cryogenic system in a massive three-story building. Thus the device, which weighed eighty-two tons, was more an industrial plant than a weapon.

The early-morning blast created a fireball more than three miles wide as well as a wave of heat that was felt by observers thirty miles away. The Nagasaki A-bomb explosion was 22 kilotons; this blast was 10.4 megatons, equal to nearly five hundred Nagasakis. Elugelab Island vanished and was replaced by a mile-wide undersea crater that was sixteen stories deep.

For decades afterward, the public believed that the hydrogen bomb derived most of its explosiveness from hydrogen fusing, which would imitate the sun's secret recipe. Not so. Almost immediately, bomb designers had realized that hydrogen fusion's extreme heat and the emission of a profusion of fast neutrons could efficiently initiate fission in materials that ordinarily couldn't sustain fissioning, materials like inexpensive uranium-238, so they'd encased

The first H-bomb explosion, on November 1, 1952, used liquid hydrogen to achieve the 10.4-megaton detonation. The three-mile-wide fireball obliterated the small island of Elugelab on which the test was conducted. The Hiroshima fireball had been thirteen times smaller. *(U.S. Department of Energy)*

the bomb in natural uranium. Now they had a multistage explosive that used a fission-fusion-fission process. This has been the basic design of every nuclear weapon for the past sixty years.

The principle is simple. Ordinary uranium-238 will fission but will not produce a self-sustaining chain reaction because its fissioning does not release additional neutrons that can go out and make the next atoms release their own neutrons. U-238 will release lots of energy through fission, but *only if it's supplied with neutrons from an outside source.* Well, the hydrogen bomb's fusion process provides this outside source of fast neutrons.

According to Hans Bethe in his 1954 history of the H-bomb, Teller's original idea, referred to as Method A, was to employ a small atomic fission bomb to start the fusion process in a nearby cylindrical tank of deuterium. This, Teller thought, would produce a self-sustaining fusion reaction that would spread through the entire length of the deuterium tank to create the superbomb they were all seeking. But the best card-punch computers of the time said that the fusion would not propagate. It would quickly die out. Thus came the resurrection of the 1946 layer-cake concept of alternating deuterium and uranium with a so-called spark plug of highly enriched U-235 in the center. The hydrogen fusion would be kept propagating by a second atomic explosion from the U-235 in the center of the deuterium tank, which would come to the rescue just as the deuterium was cooling below its ignition point, thus continuing the fusioning. Along with releasing its own energy, this deuterium burning would provide a flood of fast neutrons that would then ignite and sustain fissioning in a surrounding casing of ordinary uranium. This arrangement would theoretically produce the maximum amount of power.[5]

In those giddy weapon-intoxicated days, some nuclear physicists, perhaps half, essentially said, *Stop. Enough is enough!* and advocated against proceeding to build every imaginable next generation of superbomb. Their peace protests did not prevail. Instead, spurred by the shock of the Soviet Union's first A-bomb test, on August 29, 1949, President Harry Truman ordered the super-weapon development to proceed. After the first H-bomb blast in November, which had a yield of 10.4 megatons, the general mind-set in the United States was that the nation should build bigger bombs and more of them.

There was a limit to the size of A-bombs because the start of fissioning would quickly blow apart the assembled mass of uranium or plutonium and put a halt to further explosiveness. In practice, always less than 15 percent of the precious fissile material went ka-blooey; the rest was wasted. But a hydrogen-fusion explosion could theoretically be as large as designers might wish; all they had to do was add more deuterium or tritium.

Indeed, there was a vague fear from some that hydrogen bombs would set off the hydrogen in Earth's oceans and blow up the planet, although this low-probability but high-consequence potential cataclysm *still* did not deter their development and testing. Since the devices could be made larger with virtually no limit, the actual size of technologically possible doomsday-level weapons became a real topic for debate. Theoretical physicist Robert Serber later recalled looking at Edward Teller's Los Alamos blackboard and seeing a list of ideas for weapons "with their abilities and properties displayed. For the last one on the list, the largest, the method of delivery was listed as 'Backyard.' Since that particular design would probably kill everyone on earth, there was no use carting it elsewhere."[6]

So unleashing a true cataclysm was not beyond discussion at the time (that time being the late 1940s and early 1950s). But ultimately, the final designs wouldn't be pure hydrogen bombs. Rather, most of the power would come from fissioning. Bethe described this as "Method B." The problem—and the reason it belongs in our cataclysm context—is that fissioning, unlike fusion, is extremely dirty. It produces lethal fallout that spreads around the world. Five hundred copies of the three-stage fission-fusion-fission weapon named the Mark 41 were produced, each with a yield of twenty-five megatons. It was not only the most powerful single weapon in the U.S. arsenal but also the dirtiest.

People saw this vividly during the second U.S. H-bomb test, code-named Castle Bravo, on March 1, 1954.[7]

Now for the first time, in the Castle Bravo explosion, the deuterium fusion fuel did not need to be liquefied or maintained cryogenically. Instead, the fusion material was powdered lithium deuteride at room temperature. The lithium contained two separate isotopes, and physicists wrongly believed that only one of these was capable of fusion. The detonation was supposed to be an impressive six megatons, but to everyone's surprise, the supposedly inert component, lithium-7, sustained fusion and boosted the blast to fifteen megatons, which remains the largest in U.S. testing history. The explosion produced widespread fallout that rained down on a Japanese fishing boat, the *Lucky Dragon*, sickening the crew. It was the first time the world became fully aware of the fallout dangers of nuclear weapons. Still, since the test also successfully confirmed the staging concept described earlier, the device used in this test was the basis of the nearly three hundred 7.5-ton Mark 21 thermo-nuclear bombs built during the next decade.

In the fullness of time, the United States' and Russia's nuclear stockpiles decreased, partly due to the SALT treaties and partly because of miniaturization and the use of MIRVs, or multiple war-heads on a single missile. As of 2015, the United States had 7,100 nuclear weapons and 1,890 separate means of delivery, including bombers, submarines, and MIRVs. But despite being called H-bombs, every single one uses fission for most of its explosive energy and would therefore create widespread radiation fallout.

W88 Warhead for Trident D-5 Ballistic Missile

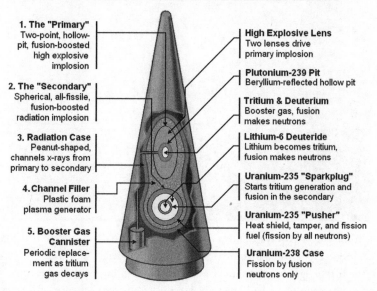

1. The "Primary"
Two-point, hollow-pit, fusion-boosted high explosive implosion

2. The "Secondary"
Spherical, all-fissile, fusion-boosted radiation implosion

3. Radiation Case
Peanut-shaped, channels x-rays from primary to secondary

4. Channel Filler
Plastic foam plasma generator

5. Booster Gas Cannister
Periodic replace-ment as tritium gas decays

High Explosive Lens
Two lenses drive primary implosion

Plutonium-239 Pit
Beryllium-reflected hollow pit

Tritium & Deuterium
Booster gas, fusion makes neutrons

Lithium-6 Deuteride
Lithium becomes tritium, fusion makes neutrons

Uranium-235 "Sparkplug"
Starts tritium generation and fusion in the secondary

Uranium-235 "Pusher"
Heat shield, tamper, and fission fuel (fission by all neutrons)

Uranium-238 Case
Fission by fusion neutrons only

The W88 is a MIRV thermonuclear warhead currently carried by U.S. subma-rines. The component labeled at the bottom right—the uranium-238 case—greatly boosts the explosive yield to 475 kilotons, or thirty times the Hiroshima bomb. This casing is largely responsible for the weapon's intense radioactive fall-out, which would deliver widespread lethality far from the target. This common feature in thermonuclear design is a big reason a nuclear war would be a global cataclysm. *(U.S. Navy)*

The same, of course, is true of the Russian arsenal. Indeed, the Russians even designed, largely for propaganda purposes, a one-hundred-megaton thermonuclear device that would have been seven times more powerful than the Castle Bravo detonation. Since it was also a staged device with an ample uranium-238 casing, it would have deeply contaminated more than ten thousand square miles, with lesser poisoning around much of Earth's Northern Hemisphere. At the last minute, they fortunately withheld much of its fissionable uranium, reducing the blast to a "mere" fifty megatons in which 97 percent of the power came from fusion alone.

Tsar Bomba, over Novaya Zemlya on October 30, 1961. This fireball is the largest explosion ever created by humans. Detonated at a height of thirteen thousand feet, the fifty-megaton blast came from a bomb weighing twenty-nine tons. The pilots of the drop plane had been given only a 50 percent chance of surviving. *(Wikipedia)*

Called Царь-бомба (meaning "Tsar-bomb" or "king of bombs"), it was detonated in the northern Siberian peninsula of Novaya Zemlya on October 30, 1961. The blast was ten times greater than all the combined armaments used in World War II. And yet it was the cleanest explosion in terms of fallout.

Unfortunately, this sort of fallout-limitation design isn't used in the actual weapons in the superpowers' inventories. A typical U.S. weapon is the W87, a three-hundred-kiloton MIRV warhead[8] of which 525 were built, and which were still in service until very recently. Like the others, it is a fission-fusion-fission device with a uranium casing, a U-235 core inside the fusion hydrogen bomb, and an aerogel foam surrounding all stages.

These five-hundred-pound W87 MIRVs are thermonuclear weapons whose manually adjustable 300- to 475-kiloton yield each has at least twenty times the power of the Hiroshima blast. Each contains an atomic bomb and a hydrogen bomb suspended in a lightweight aerogel. Riding together on a single ICBM, they are targetable with an accuracy of fifty feet. *(U.S. Government)*

The current standoff harbors several worrisome factors that would tend to make a nuclear war a nonlimiting conflagration. Unlike past wars, in which leaders could eventually declare a truce to end the fighting, a nuclear war would probably entail each side firing everything it had and holding nothing back, since withheld weaponry would presumably be destroyed by incoming missiles. Added to the global fallout—the United States estimates that six hundred million people in Russia, China, and Japan would be killed solely by fallout during a major-powers-type nuclear war— the current situation is the most perilous in human history.

But in our chilling dissection of foreseeable cataclysms in our own lifetimes or that of our children, it's arguable whether all-out nuclear war poses a more imminent threat than nuclear terrorism. After all, nearly all established nations have policies based on self-preservation. Governments possessing nuclear weapons have systems in place that are designed to make unauthorized launches difficult or impossible.

By contrast, "religious crazies" on suicide missions[9] operate under very different guiding principles. Here the notion of setting off a global conflagration may well be appealing. Self-preservation plays no role, a game-changing factor. As many have pointed out, the main stumbling block to creating a private nuclear device is obtaining uranium enriched to an extremely high U-235 content of 80 percent or plutonium-239 of similar purity. Unfortunately, countless tons of these substances are now scattered in a dozen countries around the world, and several, such as Pakistan and North Korea, have unknown but worrisome levels of commitment in terms of preventing their dissemination. Even in non-nuclear nations like Japan, large amounts of plutonium have been amassed as a by-product of nuclear power plants, and a well-executed theft is not inconceivable.

The good news is that the United States, Canada, and most European nations have put in place routine radiation monitoring in ports and even out at sea so they can intercept cargo-container-borne nuclear devices. Such neutron detectors are a giant security improvement; as recently as the 1990s, Western ports and major cities were entirely defenseless. However, other vulnerabilities to smugglers are still widely acknowledged to exist, and nobody wonders whether terrorist organizations would hesitate to destroy Western cities if they ever got the chance.

Given the vast number of religious-based radicals and the huge amount of unaccounted-for bomb-grade materials at large, it would surprise no one if a terrorist atomic-bomb attack eventually succeeded, one in which we received no warning prior to a blinding flash. People who have forsaken urban homes and chosen to relocate to rural areas sometimes cite this as their reason, even while the vast majority rarely even discuss nuclear war or terrorism.

Obviously, no consideration of cataclysms, especially those involving nuclear weapons, would be complete without at least mentioning this terrifying possibility.

CHAPTER 29

MODERN METEORS AND FLIPPING POLES

Some people are more than merely fascinated by the concept of the end of the world; they seem almost attracted by it. Maybe it's simply that the endless stream of existential threats, from North Korea's nuclear weapons to the evils of processed sugar, is exhausting. Americans and Brits are both action-oriented; let's get it over with! For some, the appeal of a final cataclysm is no doubt linked to personal problems like depression or a bad FICO score. Others may be drawn to the biblical concept of the apocalypse, with its promise of a cleansing followed by a rebirth or resurrection. Whatever the source is, people's absorption with Armageddon is widespread and never-ending.

Deadly Meteors

The author is a longtime host and on-air commentator for Slooh, the online community observatory. Visitors to the site can join one of Slooh's frequent real-time programs that use giant telescopes to stream true-color images of lunar eclipses, planet oppositions, and, most popular of all, close approaches of newly discovered near-Earth asteroids (NEAs).

Such NEAs zoom past our planet a few times a year. They're typically discovered just a few days before they come closer to the Earth than any planet and sometimes even within the orbit of the moon. On February 15, 2013, we at Slooh announced that we would offer real-time telescope observations of an asteroid predicted to pass unusually close later that day. Then, that very morning, by sheer coincidence, a different asteroid approached from the opposite direction, the northeast, emerging from the glare of the sun. This one did not pass by us.

Residents of Chelyabinsk, Siberia, a region long kept hush-hush because it was involved with the Soviet and then the Russian nuclear-weapons programs, were startled by a brilliant "second sun" that streaked across the early-morning heavens. It left a smoke trail in the cloudless sky and exploded in a terrifying fireball eighteen and a half miles above the town.

No one would have been hurt had it not been for normal human curiosity; it drove thousands to peer out their windows at the lingering trail of white dust, and when the shock wave hit two minutes later, it violently blew out the glass windows, sending over a thousand bleeding people to the town's hospital. Later calculations showed that the culprit was a sixty-foot-wide meteor that streaked toward the Earth at a speed of twelve miles per second. It broke apart eighteen and a half miles above the ground with the violence of a 440-kiloton nuclear weapon (the power of about thirty Hiroshima bombs). It then rained valuable fragments of itself onto the snowy countryside, fragments that were quickly hunted down and collected. A few of these can still be purchased online.

This celestial interloper promptly conjured images of its dinosaur-ending cousin that smashed into the planet sixty-six million years ago, and it made everyone vividly aware that it could happen again, this time wiping out every human instead of every raptor. The Chelyabinsk meteor wounded more people than any other space object ever had, although this was not difficult, since the previous record of meteoric injury involved only one person.

That earlier event happened in the United States in 1954 and was famous enough to make the cover of *Life* magazine. To merely state that its victim, Ann Hodges, was the only person in history to be hit by a meteorite is to give short shrift to a very unusual story.

On November 30, 1954, the Sylacauga, Alabama, woman, not feeling well, fell asleep on the living-room sofa in her rented white house across the street from the Comet drive-in theater (whose neon logo depicted a zooming meteor-like object, proving once again that irony is everywhere). Hodges was snapped out of a dream when the grapefruit-size stone crashed through the living-room ceiling. Before she could begin to react, it bounced off a wooden console radio, struck her left thigh, and bruised her left hand. The incident quickly drew crowds of journalists and put the thirty-four-year-old woman in the history books, along with area physician Moody Jacobs, who became the only doctor ever to treat someone injured by an object from outer space.

But Ann Hodges did not gain any financial benefit from this historic event, unlike Michelle Knapp of Peekskill, New York, who in 1992 was paid $66,000 by a collector for a twelve-pound meteorite that smashed into her Chevy Malibu in front of her rented house. In Ann Hodges's incident, and to her incredulity, police officers and government officials took away the meteorite without her permission.

Ann and her husband hired a lawyer who eventually secured the meteorite's return, but their hopes of making a fortune from the stone quickly faded when the landlady, Birdie Guy, claimed the meteorite was rightfully hers and fought for custody of it in court. The Hodgeses finally settled the case by giving their landlady five hundred dollars for the space rock, but by then the headlines were long over and the invader was no longer a hot or valuable item. The couple eventually turned it over to the Alabama Museum of Natural History at the University of Alabama in Tuscaloosa, where it remains on display.

Ann Hodges's bruise was the only authenticated human injury

from a hurtling space object for the next fifty-nine years, until Chelyabinsk's flurry of cuts from window glass.

It was all more than enough to keep alive the peril-from-space motif—the fear of a giant Earth-smiting stone. In 1996, *Discover* magazine inflamed that concern with an article supposedly proving that one is six times more likely to die from a meteor impact than from a commercial plane crash. That seemingly implausible figure actually makes sense, since air travel is extraordinarily safe, while a biosphere-destroying giant-meteor impact, though vanishingly unlikely in any given year, could kill half of the Earth's humans if it ever happened.

Which brings us back to those periodic Slooh programs that telescopically peer at NEAs as they whiz through Earth's neighborhood. Public comments, always welcome during the shows, feature a repetitive theme: *I don't trust our government to tell us the truth if a giant asteroid was actually on a collision course with Earth. But I do trust you guys, so that's why I come to this website.*

While our company certainly welcomes visitors, it's odd that so many think the U.S. government would keep an impending cataclysm a secret. A potentially catastrophic celestial collision certainly would *not* be kept hush-hush, for a couple of reasons. First, I count several professional astronomers among my friends, and I know them all to be blabbermouths. If any of them discovered a new asteroid, charted its predicted orbit, and found that it was going to intersect Earth's path, the fact would be immediately blabbed to colleagues, family, friends, the media, and maybe even strangers on the street. And second, the federal government does not have any sort of mechanism or protocol in place for keeping celestial discoveries a secret. There isn't even an existing national censorship mechanism for media stories. The author has been a contributing editor to three separate high-circulation national magazines for thirty years. Even when one of them (*Discover*) ran an article detailing the optical-resolution capabilities of the KH-11 spy satellite, no government representative asked the magazine to keep such stories out of print. It just doesn't happen.

People probably get their ideas about government secrecy from sci-fi films in which some military bigwig is informed about an impending earthly collision and invariably tells his men to "keep it a secret or the public will panic."

Of course, in the movie, the secret does get out and the public *does* panic. This pandemonium always takes the form of people running helter-skelter through city streets. Beyond the unlikelihood of such frenzied stampeding (given our overweight citizenry), there's the fact that the entire planet is about to be destroyed. It would thus seem that no earthly destination should offer any particular advantage. Why budge?

At least an impact cataclysm is astronomically possible. It's even likely, given enough time. Everyone is aware that it has happened before. When the author appeared on David Letterman's late-night talk show, the first thing the host did was hold up a black meteorite he had just been handed and ask, "So what are the chances some big meteorite has my name on it? And that we're all screwed?" He obviously thought the idea was amusing, not terrifying.

Pole Reversals

Equally humorous are several other possible future cataclysms that are popular with worriers, chief among them the idea that Earth's poles will flip and destroy all life. The pole-flipping disaster is an issue that astronomers are periodically asked about, and the author routinely fields questions about it when appearing on monthly radio call-in programs.

This is a supposed cataclysm we can have some fun with because, although it sounds ominous, it is not an actual threat.

Like the meteor-impact scenario, reversals of the poles are

events that have indeed happened to our planet before, many times. Scientists have recorded 184 magnetic-field reversals in the past eighty-three million years. But anyone wondering about pole flips has to clarify. Which poles? There are two pairs of poles, so you'll have to be specific about which of these has you worried. The poles most people are aware of are the poles of rotation—that is, the geographical poles, the poles around which Earth rotates, one of which is where Santa lives. These are the places farthest north and south. Those poles do shift annually, but by less than a city block.

But there's another set of poles, and those *can* shift rather quickly. These are the magnetic poles. Since most people who worry about pole shifts don't know which poles they're fretting about, here's a quick review.

The Earth's magnetic north pole, where magnetic-field lines dive vertically into the ground and toward which compasses point, currently sits five hundred miles from the geographic North Pole around which the planet rotates. It's just west of Ellesmere Island in northern Canada, where, except for a small contingent of the Canadian military, nobody in his or her right mind has ever lived. This spot was first located by explorer James Clark Ross in 1831. There's no buoy or marker there because the magnetic pole moves. Indeed, it's been moving faster and faster lately. Since 1900, the magnetic north pole has migrated six hundred and fifty miles. In the past forty years, the rate has accelerated from five miles a year to thirty-seven miles a year. That's sixteen feet per hour. The cause: motion in Earth's liquid iron outer core.

We know the historic direction of our magnetic poles because when volcanic lava solidifies, the iron in it aligns with Earth's magnetism. Such basaltic rocks reveal that our planet's field reverses its north–south orientation haphazardly and does it on average a few times each million years. The time period for each new direction of north or south is known as a *chron,* and a chron can last for millions of years or, as in one historic case, change twice in just fifty

233

thousand years. The average chron lasts four hundred and fifty thousand years. Some folks fear that the next flip is imminent, despite the fact that we've been living in the same chron for the past seven hundred and eighty thousand years. There's no reason it should happen this century, even though Earth's magnetism has weakened mysteriously by 10 percent since 1850.

But just as important, such magnetic-pole reversals have never matched the times of mass animal extinctions, so obviously, they don't hurt anyone. And the process of reversal usually requires somewhere between a thousand and ten thousand years, so it's not anything that can happen over a cup of coffee or even over a decade, although the most recent chron change appears to have completed its switch in under a century, which is extraordinarily rapid. Since then, people have grown accustomed to compass needles pointing where they presently do, to the north.

But something *very* weird did happen just forty-one thousand years ago, during the last ice age. The poles flipped completely, but the change lasted only four hundred and forty years and then returned back to the way things were and are now — and while the poles were reversing, Earth's overall magnetic field dropped to just one-twentieth of its normal strength. (Experts still regard the current chron as having lasted seven hundred and eighty thousand years, despite this quick back-and-forth shift.)

Bizarre. But mostly harmless. The fossil record shows no mass extinctions or biosphere troubles during that time, other than the fact that it was an ice age, so it was freezing cold.

As to the geographic poles, those other poles, they never shift suddenly, but they do continuously migrate in a pivot called the Chandler wobble, moving up to a hundred feet every 433 days or so. The poles more or less return to where they started. A big earthquake can throw them off by a few feet (and speed up our daily spin by a millionth of a second), but the main causes of migrating poles are temperature, salinity, and pressure changes deep in the oceans.

Since the geographical poles roam, so must the equator.

In 2007, your author visited the Ecuadoran government's equator monument outside the capital city of Quito. Beloved by tourists, the place features an enormous four-story obelisk with an equator line marked by stones and earsplitting salsa music blaring from speakers 24-7. Unfortunately, thanks to sloppy surveying, the monument is in the wrong place. A quarter of a mile away stands a popular private museum that claims to be the real GPS equator, and it has its own equator line. But the director of that museum, after learning he was being interviewed by an American science journalist, admitted that that line was not accurate either. The true equator, he insisted, was on a vacant lot a couple of hundred yards north.

But even if someone painted a third line there, it would have to be as broad as a California freeway in order to enclose the true, ever-shifting equator that migrates thanks to the Chandler wobble.

So, sure, worry about the poles flipping if you have nothing better to fret about. But first you have to find them.

CHAPTER 30

WHEN IT SURE SEEMED LIKE
THE APOCALYPSE

To primitive peoples, any natural event that was unprecedented and visually spectacular, especially if it was rife with rapid or especially frenzied movement, seemed to be the onset of the apocalypse. But some celestial alterations were worse than others. You didn't need to be the tribal elder to know that the sun's disappearance was a bad thing. A total solar eclipse, which happens in any given location about once every three hundred and sixty years on average, was just such a calamity. For most early societies, such dramatic sky configurations were more than mere visual oddities; they had meaning. Interpretation of celestial events was a big part of communal life, even if only trusted village elders or priests usually made such sky-based assessments. To one and all, the sun being blotted out in the middle of the day did not bode well for the future.

Still, one may wonder whether some perspicacious tribe members figured out what was going on and reacted with awe instead of terror. Following the moon's motion in its unvarying 27.32166-day circuit around the sky or its changing phases that obey a separate 29.5-day synodic cycle, a keen observer would certainly know that the moon was approaching the sun's position in the sky and might still see it when it was a hair-thin crescent low in twilight and only about fifteen degrees from the solar disk. He or she might confidently predict when these two most important bodies would meet in the sky. If an eclipse happened at around the expected time of juxtaposition, wouldn't that observer, priest, local genius, or village astronomer say to his or her family, "That's merely the moon in front of the sun. Look, it's still in motion. Don't worry; it will keep moving on."

The Babylonian astronomers possessed such abilities. These sky watchers even predicted eclipses decades in the future. In our modern era, computer programs spell out the moon's shadow's location to within a fraction of a mile centuries in advance.

So it's odd that we still get taken by surprise and then imagine the worst.

When the Stars Fell on Alabama

The Leonids certainly surprised people. In 1799, 1833, and 1866, this normally skimpy meteor display did something bewilderingly spectacular. In the predawn hours of November 18 in each of those years, observers saw sixty meteors *per second*. The sky was continuously filled with streaking shooting stars, all radiating from a single spot in the heavens; it was as if someone had drilled a hole through the black firmament and sparks were crazily leaking from it.

Sixty a second. Some were brilliant fireballs. A few exploded like fireworks. Most left behind lingering glowing trails that had a green tint, all of which pointed away from a single spot near the star Algieba in the constellation of Leo the Lion. It seemed, according to several observers, like a glowing snowstorm in the clear night air. The 1833 display, visible in cloudless weather in the American South, had entire plantations of slaves holding each other while watching the sky. Some of them interpreted it as the start of a cataclysm. But others thought it was a sign from God that their hard lives would soon end and that emancipation wasn't too far in the future. The event inspired the song "Stars Fell on Alabama."

With the thirty-three-year periodicity established by 1866 and astronomers assuring everyone that Earth was merely plowing through a rich swarm of icy or dusty material shed by the Tempel-Tuttle comet, which indeed returns to the inner solar system every thirty-three years, the next expected spectacle, in 1899, would be the first that could be greeted with excitement rather than end-of-days hysteria. But November 18, 1899, came and went, and people saw only a normal meteor shower, meaning around fifteen extra shooting stars per hour. It was a yawner.

When 1933 similarly passed without any meteor storm, astronomers decided that they'd seen all they were going to see. The orbit of the comet, along with all the debris that generally fell off it like gravel from an overloaded truck, had now shifted its position. The giant planet Jupiter must have perturbed its orbit. There would be no more storms.

This is why no one was particularly interested in observing the sky on November 18, 1966. But with this as background, let's switch the scene to Texas that night, in the quiet post-midnight hours. One of the author's former students, Victoria Drowns, who had just turned twenty years old in 1966, woke up in her train compartment and idly looked out the window.

With a start, she bolted upright. The sky was exploding. In less than a minute, she realized that a cataclysm was unfolding. It was the end of the world. What else could it be? She looked around the compartment. Everyone was asleep. What should she do? These were strangers around her. What was the appropriate behavior when you opened your eyes on a train and realized the world was ending? What was the etiquette in this situation? Were you supposed to wake people up for Armageddon or let them sleep through it?

The conductor came by and she frantically waved him over to look out the window. The two of them found empty seats away from the sleepers, and together they quietly watched the fireworks until dawn erased them. Victoria was mildly surprised the next day that the world continued and, indeed, that few knew anything about it.

I heard a similar account from another student who had been a soldier on guard duty at the time. All this is shared because that night in 1966 was the most convincing cataclysm-that-wasn't in modern times. And in case you're wondering, yes, certainly everyone remotely involved in astronomy set the alarm to observe the heavens in 1999. In fact, we started watching in November of 1998 to be on the safe side. And although some places around the world reported a rich meteor shower in 1999, there was no storm.

We watched the next year, 2000, but again no dice. But as the autumn of 2001 approached, some meteor-shower experts, like NASA's Peter Jenniskens, predicted that 2001 would indeed be a year of greatly heightened activity. It so happened that the night of November 17 was cloudless, a somewhat rare bit of good fortune in the mountains of Upstate New York, where your author has lived for some forty-five years. I woke the kids and we all huddled in sleeping bags in the meadow behind our house. And there, starting around two in the morning on November 18, we saw the best meteor display of our lives. It was no storm, no case of multiple meteors per second. Nor was it merely a rich shower with perhaps one shooting star per minute. What we observed for several hours was a steady five to six bright green meteors per minute, nearly all of which left behind lingering trails that hovered in place and then slowly faded like Cheshire Cat smiles. It was wondrous.

If people hadn't known what it was, might they have interpreted it as the start of a cataclysm, as Armageddon? No, probably not. It didn't cross the line from spectacle to terrifying apparition. Meteor experts do not expect one of *those* to happen until 2099.

But events other than a fiery-looking rain of meteors can seem to herald Armageddon. Way back in 1910, the world was terrified by a predicted celestial cataclysm that hadn't yet created the slightest disturbance in the sky.

Cyanide Calamity

The year 1910 was strange from the outset. The normal frequency of great comets—defined as a comet bright enough to be seen clearly with the naked eye even from cities—is one every fifteen to twenty years. In our own lifetimes, such bright interlopers were Comet Ikeya-Seki in 1965, Comet Bennett in 1970, Comet West in 1976, Comet Hale-Bopp in 1996, and, perhaps uncounted because it was visible only to those living deep in the Southern Hemisphere, Comet McNaught in 2007. So roughly five in fifty years.

But 1910 began with an amazingly brilliant apparition later known as the Great January Comet, which was a surprise because astronomers had long predicted a return of Halley's comet in May of that same year, and this meant that two brilliant comets would parade overhead within a mere four-month interval. Extraordinary indeed.

The Great January Comet shone for two weeks, starting in the middle of that month, and had a tail fifty degrees long, or a hundred times the visible width of the moon.

But Halley was even better for those who set the alarm and looked eastward just before dawn during almost the entire month of May. Its peak was May 19, when, after the moon had set, the entire tail could be seen to stretch two-thirds of the way across the sky.

Much later in the twentieth century, non-astronomers who'd seen both comets typically conflated the two. When the author as a teenager questioned his grandmother about what she'd seen in 1910 at the age of fourteen, she "remembered" Halley's comet, but in all likelihood it was the brighter of the two interlopers, the January event, that she was actually recalling, especially since this was the only comet that was at its best in the convenient evening viewing period.

She had also forgotten that people back then did not really welcome Halley's comet. That's because soon after the spectroscope was invented, in 1859, it was used to determine the composition of heavenly bodies, and chemists and physicists saw the fingerprint of cyanogen $(CN)_2$ in comet tails. There was now a clear chemical explanation for the typical green color of a comet tail, and it seemed deeply ominous, for cyanide had recently surpassed arsenic's reputation as the deadliest of all poisons.

Then came refinements in Halley's orbit, which showed that Earth would actually brush through the comet's outer tail on the night of May 18, 1910. This was more than enough information for people to decide that the atmosphere would be poisoned by cyanide!

Astronomers tried to reassure the public by way of the media that a comet tail was so skimpy and rarefied, no one would even notice when it met Earth's far thicker atmosphere. But that had no effect. The tabloids of that era proclaimed a coming cataclysm. It would be the death of all earthly life. The public's reaction was an uncanny mixture of panic and celebration.

The weeks leading up to May 18 saw the widespread peddling of tonics and comet pills that would supposedly protect against the cyanide. Many took to storm cellars or sealed off their homes, shutting the flues in chimneys and stuffing cloth into all door cracks. They hid out the night of May 18 and tried not to breathe too deeply. But others went out to pubs and dances and partied the night away, having decided to spend their final hours in the merry company of friends and strangers.

Needless to say, there was no harmful effect whatsoever (unless you believe the comet was somehow related to the outbreak of the Great War four years later). Still, at least there was some natural, reasonable germ of truth behind the fears. This would not be the case for the three other great cataclysmic predictions that occurred during the lifetimes of many of this book's readers.

California Almost Topples into the Sea

The first prediction appeared in *The Jupiter Effect*. This was a 1974 book by astronomers John Gribbin and Stephen Plagemann that hypothesized that a 1982 arrangement of planets would cause the sun to respond with eruptions and, in an odd series of causes and effects, create earthly cataclysms that would include a major shift of the San Andreas Fault, causing unprecedented violent earthquakes and culminating in California being pulled into the sea on March 10, 1982.

In one of your author's earlier titles, a chapter contained a mild condemnation of those astronomers for using such a mockery of science to scare people into buying their books (it was a bestseller) and thereby make themselves rich. I was later castigated by one of the authors, Gribbin, who wrote me to say that he did not get rich at all and in fact had barely made any money. I don't know why that should have made me feel better about the silly, wrong prediction that had terrified millions, but it did.

In any case, the world's astronomers tried to assuage the fears of those who took the threat seriously. They pointed out that the planets wouldn't actually form any kind of line in space; they would merely all be in the same quadrant of the sky, and previous historical planet arrangements like this had not been statistically associated with major earthquakes. And in any case, the sun didn't care where the planets were situated.

Nothing helped. The world worried. Until, of course, the date came and went and California was still firmly epoxied to Nevada.

For some reason, when a predicted cataclysm fails to occur, it doesn't dissuade anyone from buying into the next one. It didn't help that the next Armageddon offered several truly unique features.

Y2K

As the year 2000 approached, everyone knew that the calendar's numerals were about to click over in a manner unseen for the past thousand years. It wasn't merely a new century but a new millennium (at least, so most people believed, despite the argument that the new millennium wouldn't actually begin until 2001). What manner of revelry could adequately capture such an occasion? Why, the last time this had occurred, it was the era of the Crusades. This was big!

Then came the trouble. There would undoubtedly have been Armageddon fears anyway, based solely on all those digits changing like a car's odometer reaching 100,000 just as you're pulling up to the door. But now suddenly the geeks and techies joined the nervous chorus, something out of character for that particular demographic. The issue this time wasn't planets but computers. Someone realized that software codes had been representing the current year as 99 rather than 1999. In that earlier epoch of the 1990s, every byte of storage and memory was costly, an MS-DOS 386 processor was an advanced machine, Apple was struggling to convince potential purchasers to use mouse clicks instead of typing "go to" commands, and people used dial-up modems to access the web. Why waste the precious extra bytes on a four-digit year code? If the Department of Motor Vehicles had 01-24-88 on some document, could the computer mistakenly process it as *1888*? Two digits seemed adequate then. But now, suddenly, came the fear that computers wouldn't recognize the year 2000 if it was represented as 00, that the digits would be interpreted by the computers as meaning the year 1900. And then what?

The faceless army of computer programmers who normally live among us in silence and obscurity were asked by reporters for their judgments of the peril, and their predictions were not reassuring. *As soon as the silicon entities realize that a big mistake has unfolded,* the general technological opinion went, *all manner of techno-smoke will probably arise, followed by crashes and shutdowns. And since everything from bank accounts to airline systems is computer-controlled, the planet's electronics will freeze or at least mess up, and planes might even start falling from the sky.*

It sounded like chaos. Food and fuel deliveries would cease, since automated distribution centers would no longer dispatch the correct trucks or even know which routes should be taken—after all, they would think they were now suddenly delivering stuff in the year 1900. Modern life would grind to a halt. And the pandemonium would all commence at 12:00:01 on January 1, 2000.

Well? Didn't this sort of make sense?

It sure did to the author's ex-wife, who'd already, a few years earlier, given up her centuries-old estate in Cannes on the Côte d'Azur to live in the extended winters of mountainous Upstate New York, where black bears would frighteningly walk past our kitchen windows a few times a month. Life to her was already primitive and barely tolerable. Convinced that Y2K meant she'd no longer get fuel-oil deliveries, she had a twenty-five-hundred-dollar backup heating system installed. It burned kerosene, which she stored in a large, expensive new tank. She could scarcely afford any of this, but she did it anyway. She'd been influenced by Spock.

Specifically, it was a TV science documentary in which the narrator, Leonard Nimoy (who played Spock on *Star Trek*), earnestly if somewhat dispassionately emphasized the gravity of the Y2K peril and predicted a global cataclysm. In response to the uproar that provoked, some corporate executives hired computer experts to revamp their systems. Others shrugged and crossed their fingers.

The likelihood of an upcoming disaster gained enormous global attention at least partially because the threat had a popular name. When it comes to high-tech or celestial hazards, a short catchy label is essential. Black holes would never have gained their cachet and vague aura of menace if they'd been stuck with their original designation, *frozen stars*. Physicist John Wheeler coined the term *black hole* in 1967, and it caught on immediately, just like *wormhole*, his previous creation, had. So now *millennial bug* was replaced overnight with the snappy *Y2K*. As *Time* magazine explained in an apocalyptic 1999 cover story, "The bug at the center of the Year 2000 mess is fairly simple. In what's proving to be a ludicrously shortsighted shortcut, many system programmers set aside only two digits to denote the year in dates, as in 06/15/98 rather than 06/15/1998. Trouble is, when the computer's clock strikes 2000, the math can get screwy. Date-based equations like 98 − 97 = 1 become 00 − 97 = −97. That can prompt some computers to do the wrong thing and stop others from doing anything at all."

The days leading to the predicted cataclysm grew more frenzied. Finally the dreaded midnight arrived, and a surprising thing happened. It was an outcome no one had anticipated.

Nothing. Nothing happened. And "nothing" was indeed amazing, because you'd think, given all the thousands of businesses and hospitals and planes and all, *something* should have gone wrong. But nothing did. The year became 00 and life went on.

The 2012 End-of-Days

And that, you'd imagine, should have ended any calendar-based cataclysm scare for a long time. But no such luck. A mere decade later, someone figured that the Mayan calendar, of all things, reached the end of one of its esoteric cycles, the so-called long count, on the winter solstice of 2012. So now *this* would be Armageddon.

Western civilization thus reverted to an unusually silly end-of-days fueled by inane apocalypse-prediction books. After the actual but negligible peril of Y2K, the world was now somehow okay with a wholly imaginary threat. The Mayan civilization suddenly gained global attention. A culture that couldn't even foresee the Spanish invaders who would soon obliterate them were now credited with transcendent prophetic abilities extending half a millennium into their future. Your author agreed to conduct a sold-out tour in Maya-land for clients suddenly psyched to see the Mayan pyramids, hike through Guatemalan jungles, and hear lectures about Mayan timekeeping. The author ran into the son of an old Mexican tour-guide friend, Moisés Morales, and he confided that absolutely no one who lived in Chiapas had the slightest concern about the world ending. However, they all looked forward to the arrival of a flood of wealthy *norteamericanos*.

In places where citizens were perennially attracted to New Agey paranoia, like the author's beloved Woodstock, the attention given to this winter-solstice event was intense. Not wishing to be left out, Hollywood got in on the action with the big-budget disaster movie *2012* starring John Cusack, Woody Harrelson, and an actor with the hard-to-pronounce name of Chiwetel Ejiofor. Its tidal waves and earthquakes looked very cool.

But when the solstice brought nothing worse than the onset of winter, it seemed we were out of cataclysms for a while. And if you believe that, you'll believe that the next cataclysm involves the collapse of the Brooklyn Bridge. Turns out, some of the most compelling global perils saved themselves for the twenty-first century, and we'll visit them in chapter 33 when we gaze with paranoid thoroughness into our always-precarious future.

CHAPTER 31

INVADING OUR BODIES TODAY

Some of the cataclysms we have explored occurred on Earth. Some unfolded near us in space. But a faraway cataclysm that wreaks havoc in its own neighborhood and is too distant for us to even see wouldn't affect us, right?

Not quite.

Sometimes the violence is so extreme, it sends dangerous debris flying at speeds that are off-the-chart high, shrapnel that can travel all the way to our planet and even penetrate our bodies to kill some of us. Impossible? Well, gather round the campfire and hear about a peril unmentioned in your insurance policy's fine print.

The bullet fragments are cosmic rays, arguably the most intimidating label in all of science. They definitely live up to it. Even their history is intriguing.

Their presence was first suspected over a century ago, when physicists noticed that some electrified laboratory objects slowly lost their positive or negative charge for no apparent reason. Some electrical entity was apparently sneaking into the lab. That it came from the sky was not at all obvious until 1912, when Austrian physicist Victor Hess carried a charge-measuring instrument aloft in a balloon and found that charges increased as he ascended. The culprit was assumed to be some kind of invisible beam from outer space, and thus the term *cosmic ray* was born.

It took until 1950 for scientists to prove that these were not rays—meaning electromagnetic beams—but solid particles, although the original name remains. Further studies showed that 89 percent of them were ordinary protons while 10 percent were alpha particles—packages of two protons and two neutrons. The remaining 1 percent were electrons. All carried electric charges and were thus influenced by magnetic fields.

A proton is a hydrogen's nucleus. An alpha particle (two protons and two neutrons) is a helium's nucleus. Their cosmic-ray proportions match the relative ratio of hydrogen to helium in the cosmos. It's as if pieces of the universe are leaking into our bedrooms. What doesn't make sense is why electrons—more abundant than protons and alpha particles in the universe—are so underrepresented in cosmic rays. But that's only the first of the mysteries.

A cosmic ray's energy depends on its speed, since a faster-moving object does more damage than a slower one. Cosmic rays (CRs) arrive in an amazing range of velocities; their power is expressed in electron volts, or eVs. The commonest, slower ones were created in the sun and present little mystery. The higher-energy CRs come from deep space. Then there are ultrahigh-energy cosmic rays, or UHECRs, which are so incredibly powerful that they're utterly baffling.

Humans, happily, are protected from most cosmic rays by our planet's atmosphere and, to a lesser extent, by its magnetic field. Still, enough CRs reach your body to deliver about twenty-six millirem of radiation exposure annually. It's probably not harmful and possibly even beneficial, even if it's 10 percent of a dose that would definitely sicken you.

You get an extra five millirem for every one thousand feet above sea level that your home is located. CRs really crank up their intensity in the upper troposphere at 35,000 to 42,000 feet, which is why you receive an extra millirem for each thousand miles you travel by plane. It's frequent-flier radiation. Thanks to their extended time in that high-CR environment, airline crews have a 1 percent higher lifetime cancer incidence—twenty-three cases in one hundred, instead of twenty-two.

Astronauts who leave our planet's protective magnetic field to venture to the moon or, maybe someday, to Mars face a fearsome cosmic-ray environment. In space, five thousand cosmic rays tear through the body each second. For astronauts on a multiyear mission, this will probably create an enormously elevated risk of cancer and the wholesale destruction of brain neurons. Future Mars explorers had better start out with high IQs because they might have trouble with graduate-school entrance exams upon their return.

Shielding is problematic. To achieve the same cosmic-ray blockage that Earth's atmosphere provides, you'd need to huddle beneath sixteen feet of water or something with its mass equivalent.

While most cosmic rays have energies between ten million and ten billion electron volts, the much more energetic ones intermingled with them are utterly inexplicable. Cosmic rays of over one hundred million trillion eVs have been detected periodically since 1991, and these are forty million times more powerful than anything we can create in a particle accelerator. A single such cosmic-ray particle can deliver a wallop equal to a tennis ball hitting you at a hundred miles an hour, even though it's far smaller than an atom. They're assumed to be protons traveling at just under the speed of light.

How does a proton with its substantial mass get accelerated that crazily? No known process can do it. For years, the leading candidates for such UHECRs have been supernova remnants, but that can't explain truly ultrahigh-energy particles. Recently, colliding galaxies like the Antennae mentioned in chapter 8 have gained favor as the source, but there are problems with this theory too. Today, the leading candidates are active galactic nuclei (AGN). They're inside the most massive galaxies, like the superheavy colossus named M87 that lies in the center of the Virgo cluster fifty million light-years from us. It's assumed these galaxies' supermassive black holes play a pivotal role in slingshotting these bullets to their fantastic speed and power. Recent measurements indicate a link between the direction of UHECRs and the location of AGN galaxies, although these are so nearly ubiquitous, it's hard to say for sure.

Perhaps the most intriguing idea is that UHECRs materialize when theoretical dark-matter particles hypothetically decay into high-speed proton pairs, one of which falls into a black hole while the other is shot across the cosmos. But is this a case where desperate, baffled astronomers are using the speculative as evidence for the exotic?

No matter which theory is correct, the origin must be some sort of cataclysmic event. That's because ultra-extreme conditions are needed to produce the required ultra-velocity. In the case of a black hole, you'd start with a star much more massive than the sun — probably weighing between ten and twenty suns — that runs out of its nuclear fuel in its old age. We saw this same scenario in chapter 5 when we explored type 2 supernovas, since this is exactly the initial start-up conditions that cause them. More precisely, we begin with a star that weighs more than eight suns but less than forty or fifty. Such superheavy stars make up less than 0.1 percent of the Milky Way's stellar inventory. The aging, heavy star, its wild youthful years behind it, is running out of fuel in its core, and loses its fierce outward-pushing energy. Now each day is worrisomely spent with all its massive outer layers hovering like a sword of Damocles, barely in equilibrium with the energy emitted by its core. Then one day, the core falters ever so slightly, and that's the final straw. The star collapses.

Collapse creates heat, so the fresh sizzle can ignite its remaining core material to perform new kinds of nuclear reactions, and the star gets a life extension. But eventually these longevity boosts run out. If the star weighs between half a sun and about 1.44 suns, it will collapse to become a white dwarf. These are common stars, which is why we observe little Earth-size white dwarfs all around us, even through backyard telescopes.

If the star is heavier than 1.44 solar masses, the stronger gravity forces a greater collapse; that can result in a type 2 supernova, which leaves behind its tiny packed core as a bizarre neutron star. We explored these in chapter 6 and came away convinced that they're the strangest observable things in the whole universe. The star is now just twelve miles wide, and its material is so packed together that a speck the size of a poppy seed would outweigh an aircraft carrier.

Yet things can get even weirder. If a star has even a greater mass, it won't stop imploding when it's twelve miles across. Instead, it will keep getting smaller with its surface gravity growing ever stronger, the implosion truly getting out of hand and showing no sign of stopping until the speed needed to escape its gassy surface reaches 186,282 miles per second, the speed of light. At that point, not even light can escape and the star instantly grows dark. It is now a black hole, although it doesn't care that it reached that stage and still keeps collapsing smaller and smaller.

Our science fails when we ask the obvious question: When does it stop? Einstein's theory of relativity says that it continues until its density is infinite and its size is smaller than a speck. Meaning, until it occupies zero space.

But since neither infinitely high density nor existence in a zero-volume space has any physical meaning, it's reasonable to ask what *really* happens. No one knows. Some think that an unknown process must step in and halt the collapse. Maybe the "strong force" that operates at atomic-nucleus scales comes to the rescue to stop the shrinkage. It's an unanswerable mystery.

Meanwhile, it has certainly been a cataclysm for the poor star and for any life on any planets orbiting it. Not because they'd get sucked into the black hole, because actually they wouldn't. No, it's a disaster because all the light has gone out, and as far as we know, there cannot be life without an energy source.

But violence begets violence, and this black hole can cause trouble in an entirely different way. If the black hole is a member of a double-star system—and about half the suns in the galaxy are binary stars—then ordinary emissions from the normal star, which include a stellar wind analogous to the solar wind blowing from our sun (made up of a steady stream of subatomic particles like electrons and protons), can be captured by the black hole's fierce gravity.

Some of this atom-stuff is pulled into the ultradense collapsed star, the singularity at the exact center, never to escape. But if its angle of approach and speed are just right, it can instead be flung away as if by a slingshot and hurled off at superhigh speeds that can approach that of light. You guessed it—it's now a cosmic ray.

There are theoretically enough of these cataclysmic black holes in binary systems to provide a steady supply of cosmic rays. Nonetheless, astrophysicists think the most energetic cosmic rays come largely or entirely from supernovas.

We explored both types of exploding stars, so we know that the mega-violence unleashed when the star explodes is almost unimaginable. It equals the brightness of five billion suns, and with type 1, there's actually very little variation in this brilliance from supernova to supernova. But it isn't just light that's being emitted. The star has been torn to shreds, and the atoms of its body are hurled outward. In some cases, the superstrong magnetic field around the new supernova accelerates charged particles for years, decades, and centuries after the initial brilliance has vanished. It's this detritus, too, that provides the cosmic-ray flux currently zooming through the entire universe and, unfortunately, your body.

Some of the particles collide with atoms in Earth's atmosphere, typically about thirty-five miles above us, which is far higher than the flights of commercial jets. When they do so, they break apart air atoms like a rack of billiard balls to create a cascade of sub-atomic items like muons, which some physicists regard as cosmic rays in their own right.

On average, 240 muons penetrate your body per second. This nonstop 24-7 violation could be prevented only if you put massive shielding between yourself and the sky, say by living in an underground parking garage. If you're not inclined to do that, you'll keep being the dartboard.

It hardly seems fair. A cataclysm many light-years away in the form of a supernova that has destroyed its own solar system now brings the destruction into your own beloved body. And you have a right to protest; a muon isn't always harmless. Occasionally, one will strike the wrong bit of genetic material in a DNA strand. This is one reason humans, even those who buy everything from health-food stores, can be plagued by spontaneous tumors.

We may enjoy the titillation of hearing about cataclysms that are distant, or that struck our world in the remote past (like those asteroid-induced mass extinctions), or that pose a threat only in the far future (like the sun swelling up so that it engulfs our planet). But the muon/UHECR business is an ongoing reality. Caused by a distant cataclysm, it continuously presents the unlikely but non-zero risk of delivering disaster into our everyday lives.

PART III

TOMORROW'S CATACLYSMS

CHAPTER 32

COLLISION WITH ANDROMEDA

The Andromeda galaxy is hurtling toward us at sixty miles per second, although some sources say seventy. Or maybe it's the Milky Way that's moving. Hard to tell, because there are no grid lines or graph paper out there to show who's in motion. No matter—it boils down to the same thing. We're gravitationally bound. The rest of the universe may be flying apart with the red-shift as common as apple pie, but the Milky Way and Andromeda are bucking the trend. Our two galaxies have gotten a thousand miles closer together since you started reading this.

The implications are obvious. We are going to collide.

It's not going to happen too far in the future either. And one well-known U.S. astronomer, Neil deGrasse Tyson, has cited this impending collision as one of his examples of how there is no "Great Intelligence" or God or benignity behind nature, since if there were, why would things like cancer and galaxy collisions happen?

But there's a surprise lurking in this particular issue of the *We're Screwed* quarterly, something very comforting about the impending Milky Way–Andromeda collision. A twist you didn't see coming. But first, let's back up and tell the story logically.

We might begin in the feel-good, post–Great War Roaring Twenties, when the brilliant astronomer Edwin Hubble announced that the famous Andromeda Nebula was not a nebula at all, not a cloud of gas in our own neighborhood, as had always been assumed. It was, he said, an island universe, the nearest example of a separate, pinwheel-shaped city of thousands of millions of suns very much like our own Milky Way. Instantly, the universe got much, much vaster than previously believed. With today's telescopes, some one hundred and fifty billion galaxies have been photographed, each with hundreds of billions of suns and probably an equally staggering array of planets, moons, and who knows what else. All told, there are probably a trillion galaxies.

Yet the average educated person can name only one (other than our own). That single galaxy has so entered our vocabulary that it has become the galactic paradigm, the most famous by far. Yes, Andromeda.

One reason for its renown is its overhead position above America and Europe, where it's perfectly placed for observation each autumn. Probably another factor is its mellifluous name. The galaxy M87 is bigger; M82 is more violent; NGC 4565 is more striking. But none have the euphonious appeal of Andromeda, a name too lovely to forget.

Andromeda is not just the nearest spiral galaxy; it's also the largest in this whole section of the universe. You'd have to venture forty million light-years to find anything bigger. But its greatest draw may simply be that it's *there*—visible to the unaided eye.

Even the lavish Milky Way star fields that spill across October's skies like blowing sand, so seemingly infinite, lie no farther away than a few thousand light-years. Among those stars but a thousand times more distant is Andromeda's ghostly glowing smudge; it could easily pass for a faint fragment of cloud, and it's utterly camouflaged by city lights.

How distant? Astronauts traveling fast enough to touch the moon in just three days would require five hundred billion years to reach Andromeda, or dozens of times longer than the universe has existed. Like snowflakes blown against a windowpane, the night's stars are mere foreground specks bearing no spatial connection to that enormous object looming in the distance.

It's big, all right. Otherwise, how could something so remote take up four degrees of our sky—as large as eight moons laid end to end? Even its brighter central section, the part seen as a glowing oval by the naked eye, appears larger than the moon. Its enormousness might be grasped by comparing distances: Moonlight reaches our eyes after traveling less than two seconds, but Andromeda's frozen portrait arrives after hurtling through space for two and a half million years. The somewhat suspect 2001 astrometric data from the Hipparcos satellite lists it as 2.9 million light-years away. Subtract nearly 10 percent for that instrument's now-infamous screwups, and it's still a whopping 2.5 million. The light we see today started out toward us before grass yet existed on our planet, providing a novel reason why no Andromedans would ever journey here seeking greener pastures.

If Andromeda's so large in our sky, it must be simply awesome through a big telescope, right? Wrong! Despite the disease of high-poweritis that temporarily afflicts most beginners, many celestial objects look best under low magnification, and Andromeda's a perfect example. The biggest problem is attaining a sufficiently *low* power so that the whole galaxy can fit in the field of view. Through most telescopes, the task is impossible. Andromeda's just too big, and only a small, nebulous, unimpressive section can be seen at a time. However, binoculars are close to perfect.

Larger instruments do brighten the image, but seeing individual stars in galaxies is tough to do. Even the giant Keck telescope in Hawaii has never visually resolved the stars in any galaxy. At my observatory, we rarely show Andromeda during public-viewing nights. It's just too anticlimactic when something so famous appears as a disappointing blur! (By contrast, the galaxy M51, a hundred times fainter, is more impressive telescopically thanks to its prominent spiral arms.)

Long-exposure photography is what's really needed for a good image of a galaxy, since film or CCDs can accumulate faint detail while the eye cannot. But here too Andromeda's a hopeless case because it's tilted only thirteen degrees from edgewise, disguising its spiral arms no matter what we do. At best, they appear as curving dust lanes, and true spirality must be inferred.

The Andromeda galaxy is oriented just thirteen degrees from edgewise. It is sixty miles closer to us each second. *(NASA/JPL)*

By sheer coincidence, our nearly edgewise view of Andromeda matches the view any Andromedans would have of us. The fact that Andromeda appears near the Milky Way in our night sky automatically tells us that from Andromeda, we are oriented almost sideways, as if there's a mirror between the two galaxies and each one is merely seeing itself.

It's a safe bet that intelligent life gazes upward from numerous

Andromedan planets, curious about our own smudgy galaxy. Do they realize, as we do, that our two systems are approaching each other? With the separation decreasing by sixty miles each second, there's been a growing recognized possibility of a collision for decades. But a NASA study completed in 2012 removed all doubt. Our two systems are heading directly into a full front-on impact, not a slightly glancing blow.

In four billion years or so, the fireworks begin. The issue for us is how clearly it fulfills the definition of cataclysm. After all, if the main consequence is that Andromeda ceases to exist, well, most of us can down a few beers and get over it. If the sudden increase in tidal forces creates an extreme burst of star formation, with its attendant gamma-ray production and supernova detonations, well, that will be attention-getting and certainly destructive to the unfortunate solar systems that are too close. There will be local cataclysms in parts of Andromeda and in sections of our own Milky Way.

But in general, colliding galaxies are mostly smoke and mirrors. Whenever such things happen, as we saw in our Whirlpool galaxy exploration of chapter 8, the galaxies do dramatically warp and contort their shapes. They look like an ongoing explosion, and their wreckage is twisted and perhaps ugly. But for those who live there, not much negativity unfolds.

You see, as an average throughout the cosmos, galaxies are typically separated by twenty galaxy widths. This makes interactions and even collisions far from super-rare. But stars within those galaxies are separated by an average of one million star-widths, so star collisions are rare enough that their occurrence generates research papers.

On a scale model, if our planet Earth is represented by a microscopic germ, then the sun, an inch away, would be the size of the period at the end of this sentence. And the nearest star to the sun and germ-Earth would be another period at the end of a sentence in a book someone's reading four miles away. Dots or periods separated by four miles. You can see that collisions would be very unlikely.

As the galaxies collide, their constituent stars are thus too far apart for individual contact. Some, however, will be gravitationally yanked from one galaxy to the other, offering us the attractive prospect of someday switching allegiance and joining Andromeda. But once the actual collision happens, the concept of switching loses its meaning because a single new galaxy ultimately forms. Right away, the new entity destroys the lovely spiral formations of both the Milky Way and Andromeda—a catastrophe in an aesthetic sense. The whirling spiral arms and glowing red and blue clouds of nebulae average themselves into smears of motion that will require two billion years to achieve a new stable configuration.

But when the smoke clears, in this spot will hover an enormous new galaxy with a spherical shape resembling some of the "giant elliptical" galaxies that are now seen in the constellations of Virgo and adjacent Coma Berenices. Sitting heavily, like Jabba the Hutt, and ruling this section of space with its awesome gravity, it may even be the target of an additional merger for the galaxy M33, a smallish blue-colored spiral currently in the faint constellation of Triangulum.

The Milky Way has, over the eons, swallowed up a few dwarf irregular galaxies; the remnant of one is the strangely huge globular cluster Omega Centauri. Thus our Milky Way gets its own payback and is swallowed in the collision, with the Triangulum M33 galaxy thrown in just to add even more mass to the new monster sphere.

No doubt, all these collisions will end up producing a supermassive black hole at the heart of it all. This would be a big change. *Nowadays,* our Milky Way is one of the few galaxies with a relatively low-mass black hole at its heart; ours, with the uninspired name of Sagittarius A* (pronounced "Sagittarius A-star"), has the mass of only four million suns. But our galaxy will become part of the new Milkomeda galaxy (yes, it's already been named, although it will be billions of years before that name can be affixed to the actual object).

And our new central supermassive black hole? If it's typical of giant galaxies, it will have strange dual jets beaming light-speed energies in opposite directions. This, of course, will present its own perils for anything in the vicinity.

In short, cataclysms will certainly unfold. But they will be consequences, or afterthoughts. In most respects, the collision and destruction of our own beloved Milky Way will be no catastrophe on its own.

A hypothetical view far from city lights starts off (upper left) with today's view of our own Milky Way, seen from our interior, worm's-eye position on Earth, with the Andromeda galaxy 2.5 million light-years away. As it approaches nearer, both galaxies eventually distort because of tidal stresses, and both galaxies combine into a new spherical galaxy. Ultimately, our view looks toward the combined core of the new supermassive galaxy Milkomeda. *(NASA/ESA)*

It all offers enough of a trailer, a preview, to make any curious person want to get away from the city lights and look directly overhead on a moonless October night at eleven, a November night at nine, or an early-December night at seven. For there, straight up — and coming directly at you as if you were watching a 3-D movie — is the universe's only easy-naked-eye galaxy, our eternal companion Andromeda.

CHAPTER 33

UPCOMING CATACLYSMS

They're not over. They're never over. As long as several major religions predict that the world will end not with a whimper but a bang, we'll have millions watching for signs. As we saw in chapter 30, it doesn't take much. The apprehension can begin with nothing more exciting than the upcoming arrival of a naked-eye comet. If you look up the Heaven's Gate religious cult in a popular information source like Wikipedia, you can only sadly shake your head:

> Heaven's Gate was an American UFO religious millenarian group based in San Diego, California, founded in 1974 and led by Marshall Applewhite (1931–1997) and Bonnie Nettles (1927–1985). On March 26, 1997, police discovered the bodies of thirty-nine members of the group, who had participated in a mass suicide in order to reach what they believed was an extraterrestrial spacecraft following Comet Hale-Bopp.

What, then, will happen during the next dramatic celestial event? We've seen that the 1910 Halley's comet visit brought a widespread doomsday fear that cyanide in the tail would poison the air. And that ninety years later, the public widely feared that a software deficiency would create global disaster. In this manner, predicted future celestial events rarely carry any actual danger, but

265

it doesn't matter, since cataclysm hysteria overcomes millions or billions even in the total absence of a plausible peril.

The sense of panic often arises with the tiniest trigger. So maybe we'd better see what's coming up in the heavens. This way, we can reliably predict when some of our neighbors might start getting twitchy and wide-eyed.

We won't have long to wait.

Let's itemize the definite upcoming celestial headline-makers.

The Great Conjunction of December 21, 2020

This will receive global headlines, and its media appeal will be boosted because the event happens smack-dab on the solstice, just like the 2012 Mayan calendar scare. Count on this one stirring up the Armageddon crowd. Truth be told, the author is very excited about it too. This will be the closest planet conjunction of our lives. And it will be observable, unlike most close encounters between planets, which happen in the sun's glare and are hopelessly invisible.

This involves Jupiter and Saturn, the largest planets and arguably the most photogenic. Right after sunset, somewhat low in the west but not too close to the horizon, the giant worlds will seem to merge into a single, very bright "star," at least for those who skipped their last optometrist appointment. They'll certainly be so close to each other that they'll appear together in the same high-power telescope field. Truly spectacular.

In fact, the author's computer simulation depicts Saturn as hovering as close to Jupiter that evening as one of its own moons did a night earlier, at a right angle to its lined-up satellites. It'll look amazing. Visually a knockout, and it'll be historically amazing too, because one sees such an extremely tight meeting of planets only once in a lifetime.

In physical space, however, the planets won't be anywhere near each other. Saturn will be twice as far from us as Jupiter is. If you drew a straight line between us and Saturn, Jupiter would hover at about the halfway point. But those who follow astrology don't care that this is a clean miss in physical space. For them, the two planets appearing together is all that matters. And it's particularly propitious because to most sky-watching civilizations, from the Chinese to the Mayan, a Jupiter-Saturn meeting was always an event of great importance and solemnity. It's the rarest of the conjunctions between bright planets, and it happens only every twenty years. Indeed, on the internet you can find all sorts of seeming correspondences between Jupiter-Saturn meetings and earthly events such as presidential assassinations. The bi-decadal meeting of Jupiter and Saturn has long been termed a great conjunction. That's *great* as in Great Lakes and great blue heron, not *great* as in "great you could drop by."

That this 2020 greatness will be the closest-together great conjunction in more than a century will drive millions into a frenzy— count on it. What may also matter is that the planets meet in the constellation of Capricornus the Sea Goat. Many will regard this as a sign of doom, although casual backyard sky-gazers will merely be confused, because what the heck is a sea goat, anyway? Adding to the potential confusion is the fact that astrologers do not care about the actual star patterns of the constellations. Instead, they use imaginary signs that do not correspond with the night's stars, and by this reckoning, this great conjunction will actually happen in the sign of Aquarius, which is the imaginary figure superimposed on the real constellation of Capricornus. I hope you're taking notes on all this.

So it will be your choice that December evening. Either stock up on canned goods and bottled water and prepare for the apocalypse by diving for your storm cellar or head in the opposite direction and haul out your telescope for the most amazing planet view of your life.

The 2027 Darkness over Egypt

Through most of history, total solar eclipses terrified the population into believing the apocalypse had arrived. In recent times, this traditional reaction seems to have subsided in favor of global publicity, science focus, and feverish travel preparations by eclipse chasers. The widespread media attention leading up to the U.S. totality of August 21, 2017, produced no accompanying end-of-world internet activity, and the extraordinary total solar eclipse on August 2, 2027, may similarly arrive with no panic.

However, this particular "darkness at noon" will happen directly over the Nile River in the Holy Land region that commands attention in religious circles. Moreover, this will be the longest totality until the next century. More than enough to award it a yellow alert as a potential focus for pseudoscientific interest or fearsome portents.

Apophis Asteroid: Close Call in 2029

In 2004, astronomers at Arizona's Kitt Peak observatory discovered an asteroid that was three times closer to us than Venus will ever get, and it was soon named Apophis.

When they calculated the asteroid's orbit, they got a shock. On Friday the thirteenth of April in 2029, Apophis will not merely come closer to Earth than any planet. It will venture closer than even our moon, and not just by a little. In their projection, the paths of Earth and Apophis actually crossed. By December of 2004, the best calculations determined that there was a one-in-thirty-seven chance that it would collide with our world.

Apophis has a diameter of four city blocks and a speed of several miles per second, so a collision would produce a global cataclysm and maybe even a minor mass extinction like the one that most recently befell the planet sixty-six million years ago.

For weeks, it looked like the big one might really lurk in the foreseeable future. But further observations refined the orbital parameters and finally removed the possibility of an impact, although it introduced a new worry.[1]

The 2029 asteroid encounter—which will be as close to us as our own orbiting satellites—will gravitationally change the intruder's orbit. Now, after 2029, it would suddenly be very possible for it to collide during its *next* visit exactly seven years later. This new threat earned it a level-four rating on the Torino impact-hazard scale, the highest warning ever awarded to any celestial object. Attention was now focused on the Apophis visit of April 13, 2036.

For the first time in modern memory, there seemed to be an actual, reasonable chance of a global cataclysm caused by a celestial object. Was it finally going to happen? Were we toast? Had it been a bad idea to spend all that money on a kitchen renovation? Then, in August of 2006, further calculations downgraded the collision chances, and observations in 2013 finally settled the matter. There will be no collision in 2036. It will be close but no cigar.

What about Apophis after that?

The most worrisome encounter for those who are alive today will happen in the Apophis close approach of April 12, 2068. The latest observations and orbital calculations put the odds of a 2068 collision at one in one hundred and fifty thousand.

It's not likely. But neither is it as remote a possibility as winning the lottery.

As far as we can ascertain, this will be the soonest reasonable time that the planet could suffer a space-based cataclysm. And if it misses Earth, well, the game is never over. An Apophis-size object slams into Earth every eighty thousand years on average. In any event, the 2029 extremely close approach will give us all something we've never seen: a naked-eye view of an asteroid so close, it will actually visibly creep across the sky.

Halley's Comet in 2061

The most famous comet comes by every seventy-six years and, you'll recall, scared the bejesus out of the world in 1910 when Earth brushed through the outer parts of its long tail. After that, the next encounter was early in 1986, but that time the Earth was on the opposite side of the sun from the comet when it had its perihelion, its nearest solar approach point, which typically is when its tail is longest and reaches maximum brilliance, so it was probably the least favorable view of Halley since the Roman Empire.

Next time will be different. When it appears in 2061 it will be nothing short of spectacular. And that plus its fame is why the months leading up to that arrival will almost certainly find the end-of-world crowd gathering in force.

With all of these upcoming celestial spectacles, history can tell us what to expect. What usually happens is that one or two years prior, some widely shared "prophecy" links the upcoming celestial event to fantastical and frightening "coincidences," to some supposed quatrain of Nostradamus, or to some allegedly relevant and apocalyptic Bible verse, and the resulting confluence is terrifying to the naive.

I'm offering better than even odds that this will indeed happen with at least three of these four sky spectacles, and each time, it will produce a new, self-amplifying social media frenzy.

CHAPTER 34

HOLOCENE EXTINCTION

It's the most imminent catastrophe of the present time or, possibly, the near future. Alternatively, it's not a crisis and it'll all somehow work out. But we cannot ignore the burgeoning human population and the way it's gobbling up planetary resources while significant numbers of species pay the price by vanishing.

The nature of this imminent catastrophe seems to have changed for the better recently. Half a century ago, the popular ZPG (zero population growth) movement, inspired by Stanford professor Paul Ehrlich's 1968 bestseller *The Population Bomb*, warned of the imminent dangers of overpopulation. Widely publicized graphs predicted that the then-worrisome human global population of 3.6 billion would double by 2015—as indeed it has.

At the time, the greatest concern about this population explosion was looming widespread famine. Ehrlich's book flat-out declared that the citizens of India and other countries would suffer devastating mass starvation within fifteen years, meaning by 1983. It also predicted that England would cease to exist by the year 2000.

In the 1980s and 1990s, the sense of impending doom grew with a series of dire warnings about natural resources. Supposedly, "peak oil" would occur in the 1990s, after which that staple of global energy would get ever scarcer, its price skyrocketing into unaffordability. Copper, zinc, and other commodities would supposedly grow scarce too. All this seemed particularly likely when experts in the 1990s started pointing out that China's new steamrolling economy would be gobbling up ever greater shares of the finite supply of global commodities.

At the same time, experts predicted that the exploding human population would jeopardize the availability of necessities such as fresh water. And as if all this weren't enough to drive everyone to seek antidepressants, the oceans were being overfished, the rain forests cut down to make room for cattle raising, and housing was encroaching on the remaining wildlands, causing habitat destruction that was estimated to extinguish as many as one hundred animal species each and every day. It certainly looked bleak.

Many of those fears have proven overblown in the fullness of time, although some still retain their validity even now, but our job is to see whether it all constitutes a present-day cataclysm.

In a way, the story is an old one. Following Thomas Malthus's famously bleak 1798 work *An Essay on the Principle of Population,* an unrelenting parade of ominous global prognostications turned *Malthusian* into an oft-repeated adjective; nineteenth-century British economist William Jevons used it when he forecast that industrial growth would come to a stop because of a shortage of coal. Such pessimistic appraisals were widely read by the literate classes, though they never quite reached universal recognition or resulted in mass protests and demands for government action. But that changed about fifty years ago. That's when TV saturation and modern universal literacy, especially in the late 1960s and early 1970s, helped guarantee that millions would be aware of and alarmed by new predictions of imminent serious shortages of essential resources, shortages that would supposedly cause prices to surge so that global economic growth and improved living standards could not continue.

Meanwhile, the world was even more surely on the brink of oil depletion, according to experts like Marion King Hubbert, a U.S.

geologist who worked at the Shell research lab in Houston, Texas. He convinced many of the validity of the "peak oil" concept, the supposed time of maximum oil production, after which new oil fields would theoretically be harder to find and thus production would fall, resulting in sky-high prices and a drop-off in industrial production. Experts predicted the global peak-oil production would happen in 1971, then in 1989, then in 1995, and finally from 1995 to 2000.

We've all heard these things. Even if they had proved true, they still might not have resulted in a cataclysm, since that word seems inappropriate for an economic decline that might produce a radical global retreat to a lower-tech agrarian lifestyle. Even the near-simultaneous depletion of several commodities like oil and copper would arguably not be earth-shattering unless vast numbers of people were grievously harmed by it.

As it turned out, however, the gloomiest predictions were wrong. In each case, the forecast of shortages was undercut by the fundamental economic laws of supply and demand. Forecast accuracy was also hurt by the human inertial tendency to conservatively extend an existing graph line in the same direction it had recently been going, rather than courageously prognosticate a new vector. So, for example, when U.S. oil production peaked in 1985 at nine million barrels a day and then steadily declined year after year thereafter, few foresaw that rising crude prices would financially spur new innovation and novel exploration methods such as hydro-fracking and tar-sand exploitation. Or that after twenty years of decline that manifested as a bottoming-out in 2005 to 2008, U.S. oil production would start to sharply rise again and actually attain a brand-new higher all-time peak of 9,500 million daily barrels in May of 2015.

During the first Earth Day, in 1970, few imagined the world was on the doorstep of a "green revolution" of gene splicing, with its new generations of wildly productive bioengineered crops. No one envisioned the yield-boosting technologies of mechanized and even computer- and GPS-guided farming techniques that would increase harvests geometrically. And no one could have known that Ehrlich's prediction of widespread famine in India before 1983 wouldn't merely fail to materialize; as it turned out, not a single famine has hit India since his book was published.

The population graph seemed a more stubborn and gloomier item to overcome, but here, too, the trends have shown some optimistic developments. Chief among them is the global urbanization tendency; city women tend to be better educated than women in rural regions and much less inclined to regard big families as desirable. Indeed, the world's growth rate has been steadily slowing for the past twenty or twenty-five years, going from 1.55 percent in 1995 to 1.25 percent in 2005 to 1.18 percent in 2015 to 1.10 percent in 2017.

Future growth estimates have varied greatly, with the average estimates of the world population, which in 2018 was 7.4 billion, as reaching 8.6 billion in 2030, 9.8 billion in 2050, and 11.2 billion by 2100, after which it should be somewhat steady, although we are, of course, largely in the realm of the unknown.

None of this constitutes a cataclysm in itself, despite the general apocalyptic mind-set about such high population numbers a half century ago. However, the word *cataclysm* certainly applies to the effect humans are having on the world's ecosystems. The strongest negatives come from the understandable desire of all these new people for a rising standard of living, with its concomitant boosted energy use per capita. As the increasing population wants bigger homes, cars, and all the toys and gadgets of modern life, the commodities of energy, water, and food are stretched. If the required energy continues to be increasingly produced by nonpolluting renewable sources, and if vital requirements such as water can be economically recycled or derived from seawater via solar power, this energy boom needn't produce disastrous consequences.

Nonetheless, the cutting down of remaining wild places to make room for farming and improper management of marine resources will certainly create problems for the animals that share the planet. The issue of whether these reach the level of cataclysms remains to be seen. The greatest threat may be to the biodiversity of the 8.7 million species (excluding bacteria) that inhabit the world with us. As of 2018, the Millennium Ecosystem Assessment, which involved more than a thousand experts, estimated a current extinction rate of up to eighty-seven hundred species a year, or twenty-four a day. Also recently, scientists at the UN Convention on Biological Diversity released an even gloomier statement concluding that "every day, up to 150 species are lost."

But many other ecologists think that, although we're indeed losing species at a thousand to ten thousand times the background rate, this still amounts to "only" around nine hundred species per year, which is roughly three a day, or over an order of magnitude less than the more pessimistic estimates.

Claude Martin, former director of the environment group WWF International, which often publicized the danger of high extinction rates, leans more optimistically in his book *On the Edge*. For example, he cites the frightening fact that El Salvador and Puerto Rico have lost 90 percent and 99 percent of their forests, respectively. Yet the former has seen the extinction of just three of its 508 forest bird species, while Puerto Rico has lost 12 percent of its native birds.

So estimates of species extinctions vary enormously, with the controversy largely based on a debate over how many are still unknown. It's a big concern regardless. Many of these extinctions are rain-forest fauna and flora. Do any of these harbor enzymes that could cure cancer? Once a species is gone, it's gone for keeps — which is why this so-called Holocene extinction, despite our equivocation on some of the particulars, earns a check mark as a genuine cataclysm in the here and now.

CHAPTER 35

THE SUN HAS THE LAST WORD

The ultimate cataclysm? That would be our biosphere's total extinction. The obliteration of every form of life on the planet.

Such a catastrophe could not come at the hands of a global pandemic or an incoming asteroid or comet. A complete, bleach-the-Earth, biospheric wipeout couldn't even arrive if the planet physically turned on us by, say, unleashing a vast chain of ongoing volcanic eruptions resembling that of the Siberian Traps two hundred and fifty million years ago. Or if irony awakened from its long dormancy and our own so-superior intellects created our own extinction via a global thermonuclear war.

Any of these cataclysms are possible, and any could begin an hour from now.

But only one cataclysm is both absolutely certain and absolutely total.

This Armageddon-delivering item offers us safety in the present moment but spoils that good deal by promising no possible chance of it *not* happening in the future. We're talking about the sun, whose continual power output already equals ninety-six billion one-megaton hydrogen bombs detonating each and every second. That's so much energy, it's easy to imagine a transient increase or decrease of, say, a mere 2 percent. Other stars do indeed cough and burp and vary their fiery outputs by far more than this in all manner of time frames ranging from seconds to centuries.

But if the sun's internal thermostat caused it to drop its output by 2 percent so that the solar constant — the solar energy arriving on each square meter of Earth's atmosphere — fell from the current 1.36 kilowatts down to 1.33 kilowatts, this seemingly negligible decrease would cause our planet to inexorably enter a Snowball Earth condition where it would be ice- and glacier-covered from pole to pole for millions of years, and perhaps forever. That would certainly mean the death of every last land creature and probably those in the seas as well.

Conversely, if the sun's output grew by the same 2 percent, a series of intermediate terrestrial reactions or positive feedbacks would boost the effects to make a complete biospheric sterilization possible.

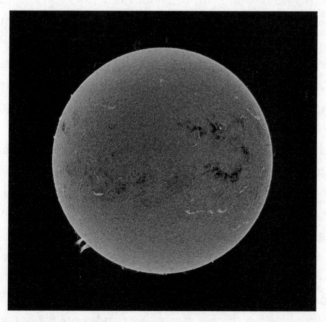

The sun's sheer size (a million times Earth's volume) and energy (four trillion watts) make any significant variations in it life-threatening to us. But major changes are coming. (*Matt Francis, Prescott Observatory*)

The sun has already delivered powerful disasters. From 1645 to 1715, its eleven-year sunspot cycle stopped cold, which resulted in widespread human suffering. Since sunspots are dark blotches on the sun, you would think that their absence during that now-famous Maunder Minimum would make the sun a bit brighter. But the opposite was true. These storms may have a black appearance when seen next to the whiteness of the rest of the solar photosphere, but they are actually brilliant and, moreover, often send shotgun blasts of broken atom fragments in our direction. These affect us. When there are many sunspots, the sun typically grows fractionally brighter in visible light and up to 15 percent more luminous in its ultraviolet emissions, which affect our planet through a series of intermediate and still not entirely understood physical mechanisms.

That period of weird sun behavior from 1645 to 1715 made for extremely cold, miserable weather on Earth. It couldn't quite be called a cataclysm, but the American colonies experienced brutal winters with widespread deaths. The canals in Venice froze. Fishing colonies in Greenland and Iceland were abandoned. It was a time of major suffering, accompanied by the odd cessation of all displays of the northern lights. Strange stuff.

Add in the powerful if unusual solar behavior of the past century, and it's obvious that the sun might be beloved, but it's not "yon constant orb" — not as steady as we might desire. Still, Earth has survived all the sun's irregularities so far. What it cannot survive is any serious future energy uptick.

Yet it is not only coming. It's already in progress. The sun is growing 10 percent more luminous every billion years. For its first four and a half billion years, Earth endured the brightening light and adapted with a thicker atmosphere and more ozone, while its inhabitants evolved to live with extra ultraviolet. But Earth cannot take much more. Climate scientists have tried to fudge their math and make this next solar increase get absorbed or go away. They've tried and failed to discover some mechanism by which our planet can compensate. About a billion years from now (a mere additional one-quarter of its already elapsed life), the sun's extra insolation will evaporate all our planet's standing water, and this will create a choking permanent cloud cover that traps in even more heat. Global temperatures will stabilize at around 710 degrees—much hotter than your oven's broil setting. At least it will be a dry heat.[1]

Fleeing to the poles won't be good enough. Perhaps some organisms will head underground, and maybe that will work for a while, deep below.

No complex life-form has endured for a billion years, and neither will *Homo bewilderus*. So while we can now speculate as to how "we humans" will handle this distressing event, "we humans" will be unrecognizable from the humans of today. Our descendants will perhaps follow the expected sci-fi route and flee to Mars, as in our first chapter's scenario. We will not be able to stay here. For no cataclysm can be more severe than the complete loss of habitat.

If you're curious about what is happening to the sun as humans flee, well, its life as a normal star, the kind astrophysicists call *main sequence*, is, at this point, only about half over. Humans may be moving farther away, but the sun is still actually quite stable in its new brighter configuration, and even after destroying Earth, it shines only somewhat more brightly; it still boasts the same size and still emits its customary white light.

Forty million more centuries then pass uneventfully. The Earth all this time is toast. The most soaring human accomplishments have long since vanished without the slightest trace. If this sounds wistfully philosophical, well, who can be blamed for mourning the passing of this entire Earth experiment, the planet whose surface swarmed with curious, quirky life-forms too numerous to have ever been fully cataloged?

Unseen all these uncounted millennia, the hydrogen in the

sun's core is consumed at an ever greater rate. But it's not enough to satisfy its new high-energy output. Increasingly it eschews this simplest food and acquires a more sophisticated palate, converting helium to carbon and oxygen. Now, four thousand million years after the death of Earth, as these processes become more dominant, the sun starts emitting so much energy that its photosphere gets shoved outward. It expands enormously, its surface stretched thinly all the way to Earth's orbit, those far-flung outer layers nearly vacuum-thin, cool, and hence orange. The sun is now a red giant.

Its width becomes a hundred times greater than before. It is now at its conspicuous peacock best. Between its gargantuan size and striking orange hue, it dominates the vistas of every planet. It's a ruddy naked-eye star visible in the skies of ten times more of the galaxy's planets than it was during the age when humans were peering back at them through their telescopes.

This new Sol persona is neither stable nor long-lived. By now the sun has endured for nearly ten billion years, but its red-giant stage doesn't last even a single billion. With a final whoosh, its core performs its last trick, a desperate move to stay alive. The alchemic changing of its new heavier elements to still heavier ones heats it up so drastically and so suddenly, it explodes.

This is not a star-destroying catastrophe. It is not a supernova or even an ordinary nova. This explosion merely blasts away the top 1 percent of its body, which rushes outward as a bubble or ring. (We see this happening to countless sun-type stars older than ours. It's so common; these distinctively symmetrical or ring-shaped gas clouds hover all around us in space. Observers in the late eighteenth century thought these strange round green blobs looked like the newly found planet Uranus and so they named them planetary nebulae. Bad term. They have nothing to do with planets.)

The sun now heats up to its maximum and turns a fierce blue. It emits torrents of sizzling ultraviolet that catches up to and excites the expanding bubble and makes its oxygen glow green and its hydrogen red. Here, ten billion years after the final episode of *Friends* and fully eight billion years after even the last reruns of it, our sun is now the central star of a planetary nebula. Its blue surface seethes at an unbelievable 180,000 degrees—almost twenty times hotter than the 11,000 degrees of the long-gone human era. No star in the galaxy is hotter than this sun. It is also shrinking, collapsing due to its own gravity. Its brief million centuries as a giant are over.

The glowing green bubble of gas surrounding it keeps rushing outward, sporting red fringes like holiday decorations. The sun's ring lasts just fifty thousand years, an *achoo* on the cosmic scale. The expanding gas then reaches a point in space too distant for the solar UV rays to excite it any longer, and the bubble fades to black. Anyway, the sun's UV is itself weakening. The superhot stage of its life, like the red-giant phase before it, has come to a close. This time, no new energy methods remain. The sun's nuclear furnace, out of fuel, sputters to an end. On the sun's eleven-billionth birthday, its pulse finally stops.

You might think the sun would then blink off abruptly like a tripped circuit breaker. However, gravitational contraction always produces heat, so the collapsing sun maintains its brilliance as it shrinks down and down, its own gravity pulling it smaller, until it brakes to a halt when it becomes, of all ironies, the size of Earth.

This new balancing point is super-stable. The sun will remain the size of Earth eternally. Its pastel-blue tint, which mimics its deceased third planet upon which it long lavished the gift of life, fades to white. It is now a white dwarf. It will shine this way for not millions but billions of years. Time means nothing to this sun. No nuclear reactions occur. No pulse, no sunspot cycle, not even any gaseous surface markings that an onlooker might use to count its rotations. Over the next billion years, it slowly cools and turns yellow, then orange, then red, then brown, and, finally, black.

The ultimate cataclysm is complete. Earth has been a lifeless, barren ball for the past five billion years, yet it continues to orbit the cold and ebony sphere that long ago was its life's provider and sustainer. Both these bodies now share the same size, though one is 330,000 times heavier than the other and almost inconceivably dense. As they perform their eternal minuet with each other, they both also keep gravitationally circling the galaxy's center every quarter billion years, like dancers sweeping around a ballroom. But they're not conspicuous. Their dual blackness camouflages them against the inkiness of space, so any future alien astronomers would be hard-pressed to detect them at all.

No further cataclysm can ever befall this inky pair of globes because neither has a future. As the early Earthling Bob Dylan sang, "When you ain't got nothing, you got nothing to lose."

All catastrophes belong solely to their past. Now and forever, nothing will change.

Acknowledgments

My thanks to John Froude, MD, a friend and infectious disease specialist, for his fact-checking and wonderful suggestions in our two disease chapters about the bubonic plague and the 1918 Spanish flu. My deepest thanks for his fact-checking also go to nuclear engineer and former U.S. nuclear submarine commander Sam Johnson.

Notes

Chapter 1: Cataclysms 101

1. In the U.S. and Britain, a telephone tone is a blend of an F and an A, frequencies of 350 and 440 hertz. In the rest of Europe, it's a strange single tone at 425 hertz, which is between an A-flat and an A and, unlike the dial tone in the U.S., is useless for tuning guitars. And in case you truly care about insect sounds, a major study showed that frequencies produced by male mosquito wings varied widely from 571 to 832 flaps per second, while females, the only ones that suck human blood, ranged more narrowly from 421 to 578 hertz, resulting in a musical note between middle A and D-flat. (Hence, people with perfect pitch should theoretically be able to assess the likelihood of whether the nearby flying pest will be landing for a snack.)

2. The pitch or tone of thunder that you hear depends on how far from you it is. A nearby lightning strike has a nearly simultaneous thunderclap, and that thunder is a fairly deafening high-pitched noise full of crisp crackles; it's not a low growl at all. The farther away the lightning, the longer the delay until you hear the thunder and the lower the thunder's pitch. Distant thunder has no high-pitched treble snaps or components; it's solely a very deep rumble. That's because low-pitched sounds travel far while high-pitched tones dissipate before they can even be carried for three seconds. (This explains why the foghorn was engineered to emit only bass notes.)

Chapter 2: It Really Was a Big Bang

1. Maybe it wasn't so effortless, since He did afterward need a full day of R & R. This may never have been a fully satisfying concept, as it merely postponed rather than solved the ultimate origins problem.

2. You'll find a disparate range of figures for "average density" in the literature, but the value given here seems among the most reasonable.

3. It was between 370,000 and 380,000 years after the Big Bang that the universe's neutral hydrogen atoms first formed, making space transparent to light. And it happened everywhere at the same time. But "the same time" doesn't mean in the snap of a finger; it took about 115,000 years before the entire cosmos was transparent everywhere.

Chapter 3: The Death of Cousin Theia

1. Back in the 1920s and early 1930s, galaxies were sometimes given the grander label *island universes.*
2. When the hydrogen isotopes are combined with oxygen, the resulting H_2O molecules are commonly called *heavy water.* It's an appropriate term because a deuterium atom has very nearly twice the mass (or weight) of a normal hydrogen atom. And a tritium atom is very nearly three times more massive. Of course, most of a water molecule's mass comes from its oxygen, because oxygen has eight protons and eight neutrons while hydrogen has just one proton. Also factoring in that each water molecule has two hydrogen atoms to just one of oxygen, the math shows that a gallon of ordinary water weighs 8.345 pounds, but a gallon of deuterium-based heavy water is 9.2 pounds. So it's a good thing Jack and Jill were asked to fetch a pail of ordinary water — probably because their dad was not involved in nuclear-weapons design like the dads we'll meet in chapter 28.

Chapter 4: Spooky Things That Went Bang

1. A truly strange naming feature of the moon involves the dark blotches so obvious to the naked eye or through binoculars, the "seas" or maria; these have all been named for either human emotions or weather phenomena. Thus we have the psychological Sea of Tranquility, Sea of Crises, and Sea of Serenity, and the weather-service-ready Sea of Clouds and Ocean of Storms.
2. According to several websites, a phobic fear of darkness is at or near the top of children's fears but ranks only around thirty-fifth among adult worries. In grown-ups, the Most Common Fears list is usually populated by arachnophobia (fear of spiders), acrophobia (fear of heights or flying), ophidiophobia (fear of snakes), agoraphobia (fear of open or crowded spaces), mysophobia (fear of germs), and cynophobia (fear of dogs, but perhaps, one suspects, not puppies).

Chapter 5: Blame It on the Supernova

1. We use supernova brightnesses in astronomy to let us calculate stuff we don't know. Let's say Jupiter and Saturn are the same size, and both are equally good at reflecting light. And both, of course, are illuminated by the same sun. So you look at both in the night sky. You whip out your light-measuring device—your photometer—and it confirms what your eyes tell you: Jupiter is much brighter. The photometer says that it's exactly six times brighter than Saturn. That sixfold brightness difference lets you use the inverse-square law (meaning you take the square root of 6, which is 2.5) and thereby conclude that if both planets are indeed equally big and reflective, Saturn must be 2.5 times more distant than Jupiter to appear six times less bright. Well, this works pretty well. It turns out Saturn is actually twice as distant as Jupiter, not 2.5 times farther, as our quick back-of-the-envelope calculations had said. Still, not too far off the mark. Turns out, Saturn is a bit smaller and a bit less shiny overall, and that's why the ringed planet is six rather than four times less bright at its perch twice as far away. Now let's do the same calculation with Uranus and Neptune, which by interesting coincidence also display a sixfold brightness difference from each other. If you're into backyard astronomy, you know that Uranus shines at a dim magnitude 6 and Neptune is an even fainter magnitude 8. Such a two-magnitude difference equals a sixfold brightness difference, since each magnitude is 2.5 times brighter or dimmer than the next. Two magnitudes means 2.5 x 2.5, and this works out to be a sixfold brightness difference between those distant blue-green worlds. Once again, the square root of 6 is 2.5, telling us that Neptune must lie between two and three times farther from us than Uranus. The approximation again works, and this time the failure to be perfectly exact is due to Uranus being slightly larger than its neighbor.

2. This type 1 supernova method of distance determination is the process used by major observatories to pinpoint the precise distances to faraway galaxies, a project intensely pursued at the Carnegie Institution's Las Campanas Observatory in Chile in its attempt to decipher the nature of dark energy.

3. Some say there are ninety-four elements, or even many more, if one includes the short-lived man-made elements. In any case, ninety elements occur in nature in appreciable amounts. Depending on whom

you ask, there are four or eight more natural elements that exist only very briefly as a result of radioactive decay of heavier elements. So the grand total of natural elements may be ninety-four or ninety-eight, but a couple are so ephemeral that you can detect them only through instruments as a brief flash of light, and you can't actually hold any in your hand. That's enough to make some of us say, "Forget it. You can't count those."

Chapter 6: Armageddon Monument

1. That's why movies of the silent era, with their sixteen frames per second, exhibited a noticeable flickering. When sound came to the cinema and the projection standard was changed to seventy-two frames per second, audiences stopped seeing an off-and-on blinking. In practice, twenty-four *different* images are shown per second but they're each displayed three times before the next picture is shown three times, and so on, yielding a rate of seventy-two frames per second.

2. Though Charles Messier made the Crab Nebula famous, and though Messier thought he'd been the first to spot it, it was actually quietly discovered twenty-seven years earlier, in 1731, by the English astronomer John Bevis.

Chapter 7: Tycho Versus Kepler

1. Hold your finger about a foot in front of your face and take turns watching it with one eye while closing the other eye. As you alternately blink each eye, you see the finger jump and appear against a different part of the background. Similarly, if you observe the moon when it's rising and again twelve hours later when it's setting, the rotating Earth has carried you as much as eight thousand miles, so you're seeing the moon from different angles. Sure enough, it appears each time against slightly different background stars. This is called parallax. If you trusted Eratosthenes's writings from two thousand years ago, knew the true size of Earth, and were armed with the knowledge of our planet's diameter, you could calculate the moon's actual distance solely by using its parallax jump. Tycho Brahe did this. And he not only easily observed the moon's parallax, which is actually twice the moon's width (or over one degree if observations were truly made on opposite sides of Earth), but also obtained the much more challenging parallaxes of the nearer planets, revealing that they were not terribly far, or at least nowhere near as

far as the then-unfathomable distances of the fixed stars, none of which showed the slightest parallax.

2. For instance, in Psalm 93: "Thou hast fixed the earth immovable."

3. This is Kepler's third law of planetary motion, and it's a cool item for everyone who enjoys a bit of simple math. Kepler noticed that if you took a planet's distance from the sun and squared it, the number would exactly equal its orbital period cubed. And these figures are conveniently expressed in Earth units. For example, sky observers since the most ancient times had seen that Jupiter required 11.86 Earth-years to move through all the constellations of the zodiac and return to its starting point. If you square this Jupiter period around the sun (that is, multiply 11.86 by 11.86), you get 140.6. This by itself means nothing, but if you cube Jupiter's distance from the sun (which is 5.2 Earth-sun units, so 5.2 x 5.2 x 5.2), you get the same 140.6! So Kepler said that by observing a planet's orbital period, squaring it, then taking the cube root of that figure, you would thereby know its distance from the sun. This was huge. And Kepler figured it out just by studying decades of records of where each planet was located.

Chapter 14: Can We Trust Space Itself?

1. Sugar cubes may seem to be a bit overemployed in these metaphors, especially since that once-ubiquitous component of teatime and coffee hour has essentially vanished. You can still purchase a box of sugar cubes at supermarkets, but they're probably not on the impulse-purchase shelves near the registers. Its usefulness in our analogies is simple: A typical sugar cube by sheer chance measures one centimeter on each side, so its volume is exactly one cubic centimeter (or cc). In the world of metric and science, when you're discussing volumes, this provides a visualization object that is almost too good to be true. A small 50 cc motorbike engine thus has cylinders whose entire ignition region equals fifty sugar cubes. It's easy to picture.

2. Neutrinos may weigh exactly nothing and would therefore travel at exactly light speed. It's a close call.

Chapter 15: The Final Supernova

1. Nothing can outrace light in the vacuum of space, but light slows down when it passes through denser transparent media, such as air, water, or glass. In those cases, objects such as neutrinos or electrons

can sometimes zoom faster than light can go through those same media. Whenever this happens and the light-speed barrier is broken, a glow termed Cherenkov radiation appears. It's the reason pools of coolant water in nuclear power plants emit an eerie blue radiance.

Chapter 17: The Oxygen Holocaust

1. That extinction event may actually not have been the most recent. The majority of modern scientists say that we are experiencing a sixth mass die-off at the present time, and they have even named it: the Holocene extinction. We'll look at this in chapter 34.
2. A eukaryote is a single-celled or multicellular organism whose cells have a cell nucleus and other organelles, each enclosed within a membrane; prokaryotes, like bacteria, don't have such structures.

Chapter 18: The Greatest Mass Extinction

1. Those who are young or new to English might find it curious to see the adjective *great* in these labels, since a mass extinction is hardly *great* in the familiar sense of "fabulous" or "cool." That's because everyday speech nowadays ignores the once-common sense of the word *great,* which was "grand" or "singularly majestic," as in the Great Plains, the Great Comet, and the Great White Way. It also means vast in scale or size.

Chapter 19: The Dino Saur Show Gets Canceled

1. L. W. Alvarez et al., "Extraterrestrial Cause for the Cretaceous-Tertiary Extinction," *Science* 208, no. 4448 (1980): 1095–1108.
2. This margin in time was originally called the Cretaceous-Tertiary, or C-T, period, and hence this was known as the C-T extinction event. But many scientists used the abbreviation *K* (from the German *Kreide,* "chalk") to refer to the Cretaceous period, so it was more often cited as the K-T event. Now that the Tertiary has been renamed the Paleogenic (Pg), the new term is *C-Pg event,* but given the longtime use of both *K-T extinction* and *C-T extinction,* those labels are still alive in some publications. (And I hope you're taking notes on all this.)
3. Go ahead, try to pronounce *Chicxulub* if you've not previously heard anyone say it or if you don't speak Mayan. Admittedly, it's not often that people chat about the impact location of the meteor that extin-

guished the dinosaurs. But if you want to be able to say it to yourself, it's pronounced "*chix*-uh-loob."

4. Bob Berman, *Zoom: How Everything Moves, from Atoms and Galaxies to Blizzards and Bees* (Boston: Little, Brown, 2014).

5. Asteroids all orbit the sun counterclockwise as seen from the north, just as the Earth does, and have orbits closely aligned with the solar system's flat ecliptic plane. These parameters tend to make earthly collisions with asteroids and their fragments (anything hitting us is called a meteor, regardless of its nature or origin) fairly low-speed events. By contrast, comets orbit the sun in random orientations and do not adhere to the ecliptic plane; the famous Halley's comet is oriented 162 degrees from the ecliptic, which is more than a right angle, and thus has a backward or retrograde orbital direction. Other comets, such as Swift-Tuttle and Tempel-Tuttle, which cause some of our most striking annual meteor showers, similarly orbit in directions opposite to Earth's motion, and thus any collision with us would be a head-on impact at around forty miles per second.

Chapter 20: Snowball Earth

1. We see sublimation in everyday life whenever we observe steam rising from a big pile of snow, such as along the periphery of a parking lot where plows have piled the white stuff. This is snow sublimating into vapor without melting into liquid first.

Chapter 21: The Plague

1. Yes, there is another side to the rat story, a much brighter one, that would be disseminated if only there were any kind of rat lobby or the slightest *Rattus rattus* public relations effort beyond the bit of industry created for rodent drug testing and breeding for the pet-store clientele. This author's wife, a longtime kindergarten teacher, always kept some manner of caged rodent in the class, and the kids adored it. One day, a principal told her, "Forget white mice, gerbils, guinea pigs, and hamsters. The best rodent by far is the pet rat. Try one." And she did. Sure enough, as advertised the rat was clean, attentive, astonishingly intelligent, loyal, and warm, and it enjoyed being held. It became adorably attached to its owners. She brought one rat home for a holiday, and I was hooked. I wrote one of my books with the rat hanging out in my

shirt pocket. I recommend these creatures unreservedly. Of course, many people can't get past the naked tail, a design flaw that dogs and cats don't have. But there's still that vestigial plague reputation. To be thought culpable for the Black Death, even if by mere association, is an insurmountable bit of biographical baggage for even advocates like your author to overcome.

2. One possible urban myth is that a famous children's rhyme has lyrics that date from the plague years and are meant to describe its terrors. But "Ring around the roses, pocket full of posies, ashes, ashes, we all fall down" is, according to musicologists, apparently only a "form a circle" play scenario, which in some variations included a kissing game and which in others pushed one person to the center of the ring; no death or dying at all. Other historians insist that the plague was indeed the origin and subject of the rhyme and game. Just thought we'd get that out of the way.

Chapter 24: Nuclear Cataclysms

1. The wisdom behind the government's orders to hastily evacuate residents in the vicinity of the nuclear power plant, an act that seemed so prudent and necessary at the time, is now seriously questioned.

2. He survived because the seven hundred rem was to an extremity, his hand. The hands and feet can take a hundred times more radiation than the rest of the body.

3. Coal dust is slightly radioactive, which is why far more people get a small annual amount of radiation from dust carried by the wind from distant coal-fired power plants than from nuclear power plants.

Chapter 25: A New State Capital?

1. TMI-1 is still cranking out 830 megawatts of electricity at the time of this writing, though it is scheduled to be shut down due to high operational costs.

2. Water hammer is a common problem in steam systems, especially when starting up the system. In a residence, it sounds like a freight train coming through your living room. In an industrial setting, the actual problem is potential damage to the turbine generator blades. The water in the water hammer is not steam; it is water vapor or droplets, and damage to the turbine blades can trash the whole turbine

because it's usually too expensive to replace a blade. If you have water hammer and you are warming up the turbines, it is very important to—hopefully with really fast controllers—shut the isolation valves for the steam pipes and throttles on the turbines.

3. This out-of-position valve was ultimately the direct cause of the accident and of the severity of damage to the core.

4. To be fair, that Ebola strain (there are hundreds of strains of it all around us continually; they're in our yards right now) killed tens of thousands in Africa, including a dozen or more American doctors who had gone over there to help as members of Doctors Without Borders. Many military doctors and nurses were sent over there too. If it had gotten a foothold in America, things would have been different. That was prevented by basically quarantining all of Africa, so the low death toll in the U.S. was not really indicative of a low-level threat.

5. Starting in the 1990s, nuclear power *was* making a strong comeback, and there was every chance that the game-changing negative publicity that had accompanied the TMI accident could be overcome. The NRC provided guidelines that made getting through the regulatory-permit process a lot easier and quicker. Power companies recognized the bureaucratic easement and committed to building new reactor power plants. Then natural gas started to flood the energy market, offering electrical generation at discount prices, and utilities started backing out of their nuclear plans.

Chapter 26: Secrets of Chernobyl

1. James Mahaffey, *Atomic Accidents* (New York: Pegasus, 2014).

Chapter 27: The Hybrid Cataclysm

1. James Mahaffey, *Atomic Accidents* (New York: Pegasus, 2014), 392.

2. Radiation hormesis is the principle that small amounts of radiation are not only not dangerous to health but possibly beneficial and confer a lifetime of cancer protection. It has much supportive evidence. But even if hormesis is not confirmed, the long-embraced but also long-criticized linear-no-threshold model, which suggests that even low doses of radiation carry some harm, is steadily being discredited, since epidemiological studies of survivors of nuclear accidents and of the atomic-bomb survivors at Hiroshima and Nagasaki continue to show far fewer cancers than that model predicted.

3. For the PDF of this projection, see http://www2.ans.org/misc/Fukushima SpecialSession-Caracappa.pdf.

Chapter 28: That Thermonuclear Business

1. In the United States, at least, the order to initiate a nuclear war reportedly requires the approval of the president plus one other person.
2. On the website of the Union of Concerned Scientists, you can find a spinning roulette wheel that details actual incidents that came close to initiating either nuclear war or the unintentional explosion of nuclear weapons (go to https://www.ucsusa.org/nuclear-weapons/close-calls# .WkpuL4IXCUk). As an example, it lists this among its thirty-six historical events:

> On January 24, 1961, a B-52 bomber carrying two 24 megaton nuclear bombs broke apart in midair, causing the pilots to eject and the bombs to drop. The force of the breakup caused the arming sequence of the bombs to start. One bomb hit the ground and shattered after its parachute failed. The other bomb's parachute deployed, with only one of the six safety switches preventing a nuclear explosion. That switch was later found to be defective in many bombs and was replaced.
>
> The crash took place in a swampy area and part of one of the bombs, containing uranium, was never recovered. The Air Force now owns an easement for the land where the bombs fell and requires permission for anyone who wants to dig in the area.
>
> After the incident, Defense Secretary Robert McNamara stated that "By the slightest margin of chance, literally the failure of two wires to cross, a nuclear explosion was averted."

3. The implosion method of creating an atomic bomb or thermonuclear blast ended up being the sole configuration used after the early 1950s, and today it is the only way the A-bomb components of such weapons are configured.
4. The public regards the very powerful 350-kiloton nuclear weapons as H-bombs or hydrogen bombs, a concept reinforced when the early superbombs were acknowledged to employ fusion involving the

hydrogen isotopes deuterium and tritium. However, those in the business of designing and building them always called them thermonuclear bombs. Actual deployed nuclear weapons all use hydrogen fusion to create a flood of fast neutrons that cause the bomb's uranium casing to fission wildly, so the vast bulk of the explosive energy comes from fission, not fusion.

5. In many modern thermonuclear weapons, the cheap U-238 is replaced with faster-to-fission U-235, since that once-scarce fuel is no longer in short supply. Using this uranium isotope helps keep the yield up and the weight down.

6. Serber's recollection is from Richard Rhodes, *Dark Sun* (New York: Simon and Schuster, 1995).

7. This particular Bikini Islands test series was called Castle. The second nuclear detonation in this sequence was Bravo, which was supposed to have a blast strength of six megatons but "ran away" from the designers and actually yielded a dangerous fifteen megatons. But sometimes the very code name for a specific nuclear detonation offers clues to its intentions. The first U.S. H-bomb test was Ivy Mike. In this case, the entire series of November 1952 Enewetak atoll nuclear explosions was code-named Ivy. *Mike,* the communications code word for the letter *M,* did not in this case indicate the thirteenth detonation in the series, as might have been inferred since *M* is the thirteenth letter of the alphabet. *Mike,* or the letter *M,* was shorthand for *megaton,* since this was the series' only explosion intended to exceed one megaton and indeed was history's first-ever nuclear explosion of such yield.

8. The W87 thermonuclear MIRV warhead actually has a dial that lets the user select the yield and choose any explosiveness between 300 and 475 kilotons; it works by limiting the amount of tritium fuel.

9. Perhaps it would be best to characterize them as lunatics masquerading as religious people.

Chapter 33: Upcoming Cataclysms

1. The reliable sequence of events is always an initial astronomical prediction of the possibility of a devastating asteroid or comet collision with Earth that's followed by, over the next few months or years, orbital refinements that downgrade or eliminate the threat. Why do further observations and data always show the danger decreasing,

never increasing? The answer is simple: There's a lot of empty space between celestial bodies. A collision is like a zooming bullet hitting another zooming bullet. Thus, if an initial calculation shows the chance of an impact, when additional information arrives, there is only a single direction that can make a collision more likely, but there are thousands of potential directional tweaks that will lead to an even greater miss. The odds are always greatly biased toward a threat reduction with the arrival of further data.

Chapter 35: The Sun Has the Last Word

1. Global cloud cover is heat-trapping for the same reason clear nights in the desert are colder than cloudy nights. This is actually the basis for how greenhouse gases work, and the fascinating process is so simple, it should be understood by all. It starts with a basic obvious fact: Daytime sun beating down on Earth's surface heats up the ground. As we all know, the objects that get hottest are dark solid materials like rooftops, asphalt parking areas, and dark rocks. Then at night, these warm materials cool off by releasing infrared energy, which heads up toward space. Here's where science enters the story. The infrared travels in a straight line through molecules in the air that have one or two atoms, which constitute over 99 percent of molecules in the atmosphere. Oxygen (O_2), nitrogen (N_2), and argon (A) all have one or two atoms, so the infrared continues straight up through them and escapes into space, cooling our planet at night.

   But here's the funny thing. Gaseous molecules with three or more atoms, like water (H_2O) and methane (CH_4), do not let the infrared (heat-produced) energy pass through easily. Instead, these absorb each IR photon and then, after a tiny fraction of a second, radiate it equally in all directions, including back downward toward the ground. So a cloud cover with all its H_2O is a place where IR from the ground is momentarily stopped and then sent radiating in all directions, and some of this actually comes back down again to reheat the ground. It's not a subtle effect. We experience this on every cloudy night, which is why TV meteorologists call a cloud cover a blanket and then raise their forecast low temps for that night.

Index

About the Author

Bob Berman, one of America's top astronomy writers, is the author of *Zapped, Zoom,* and *The Sun's Heartbeat.* He contributed the popular "Night Watchman" column to *Discover* for seventeen years and is currently a columnist for *Astronomy,* a host on Northeast Public Radio, and the astronomy editor of *The Old Farmer's Almanac.* He lives in Willow, New York.